The Ironmasters' Bags

The Ironmasters' Bags

The Postal Service in the South Wales Valleys, c.1760-c.1860

Paul Reynolds

2010

First published in 2010 by lulu.com

Copyright © Paul Reynolds 2010

The moral right of Paul Reynolds to be identified as the author of this work has been asserted in accordance with the Copyright, Designs and Patents Act of 1988.

All rights reserved. No part of this publication may be reproduced, stored in a retrieval system or transmitted in any form or by any means, electronic, mechanical, photocopying, recording or otherwise, without the prior permission of both the copyright owner and the above publisher of this book.

Every effort has been made to trace or contact all copyright holders. The publishers will be pleased to make good any omissions or rectify any mistakes brought to their attention at the earliest opportunity.

ISBN: 978-1-4457-4215-1

Front cover

Design by Antoine Cutayar, based on an engraving of c.1860 showing Commercial Place, Aberdare (by courtesy of Rhondda-Cynon-Taf library service). At this time the town's postmaster was William Morris and his printing house and post office can be seen at the end of the terrace on the left-hand side of the square.

Contents

Introduction

The intensive industrialisation of the south Wales valleys was a process that started in the middle of the eighteenth century. It resulted in an isolated and thinly populated district, largely dependent on subsistence farming, becoming one of the major economic zones of the United Kingdom. This was due to its rich reserves of iron and coal, the two products on which Britain's industrial revolution was founded. As part of the process small existing villages such as Merthyr Tydfil or Aberdare developed into towns with a population of many thousands, and completely new settlements such as Tredegar, Pontypridd or Ebbw Vale were created. The experience of the ordinary people of these towns formed one of the drivers which led to the social and political reforms which produced the society that we know today

The initial impetus for this growth was given by the iron industry. There had been a number of forges and furnaces within the valleys from the sixteenth century, but with the exception perhaps of Pontypool, they operated on a small scale. Output was limited and, as a result of topographical factors, found only a local market. The location of the works was dependent on the reserves of ore, and these lay mainly on the northern outcrop of the coalfield, twenty or more miles from the sea and navigable water. The rivers were unsuitable for navigation and the roads were of the most rudimentary character. Consequently the cost of transporting iron to the coast in any quantity and thence to a wider market was prohibitive.

In the mid-eighteenth century there was a great increase on the demand for iron for both military and domestic purposes and the potential of the south Wales valleys to meet this demand was recognised, despite the drawback of their landlocked location and the transport difficulties which this created. Iron ore was to be found in great quantities all along the northern rim of the coalfield, close to the surface and easily extracted. Just to the north of the coalfield there was a belt of limestone which could be quarried and transported fairly easily to the furnaces where it was used as a flux, an essential ingredient in the smelting process. Heavy rainfall resulted in rivers which could generally be depended upon to provide a source of power for the water-driven bellows that were needed to generate a sufficiently high temperature within the furnaces. Coal, of course, was available in abundance and this was a matter of particular importance in the growth of the south Wales iron industry, especially since the local coal had a high carbon content which kept fuel costs to a low level. The recent acceptance of coke as a smelting fuel in place of charcoal allowed full advantage to be taken of this resource. Indeed, without coke-smelting the south Wales iron industry could never have developed to the extent that it did.

The first of the coke-fired furnaces is generally believed to have been at Hirwaun, north of Aberdare, which was established in 1757.[1] It was soon followed by Dowlais (1759), Plymouth (1763) and Cyfarthfa (1765), all at Merthyr Tydfil. By 1800 a string of ironworks could be found all along the northern outcrop of the coalfield (the Heads of the Valleys or the *blaenau*) from Pontypool in the east to Ynyscedwyn in the west, including Sirhowy (1778) and Beaufort (1779), the first two coke ironworks to be erected in Monmouthshire, Penydarren (1784), Blaenavon (1789), Ebbw Vale (1790), Nantyglo (1791), Tredegar (1800), Rhymney (1800), Aberdare (1800), Abernant (1801) and others. The growth in output was rapid: in 1788, only 12,500 tons of pig iron were cast in the whole of south Wales, just half the output of Shropshire, then the leading iron district. By 1796 production topped 34,000 tons to overhaul Shropshire and by 1806 it exceeded 78,000 tons. Between 1820 and 1840 the output of iron from south Wales represented about 40 per cent of the total output of the United Kingdom. The works were built to the latest design and to unprecedented dimensions; they housed a technology of unsurpassed modernity and employed a workforce of unexampled size. Together they represented an outstanding exemplar of the latest technology and the new economic order.[2] Industrial dynasties were established, its members from one generation to another often having an interest in more than one concern. The most prominent of these were Crawshay

---

[1] However, it has been suggested that Hirwaun was originally charcoal-fired. See Riden 1993, 21
[2] Evans 1993, 28, 52

The principal ironworks of Monmouthshire

ABERGAVENNY

BRYNMAWR

Beaufort

Sirhowy •

Ebbw
Vale

• Blaenavon

Tredegar •

Nantyglo •

Union • Bute

• Blaina

• Varteg

RHYMNEY Victoria

• Abersychan

Ebwy Fawr

Ebwy Fach

R. Usk

• Pontypool

R. Sirhowy

Afon Lwyd

R. Ebwy

CAERLEON

R. Rhymney

NEWPORT

9

(Cyfarthfa, Hirwaun), Guest (Dowlais), Homfray (Penydarren, Tredegar), Hill (Plymouth), and Bailey (Beaufort, Nantyglo).[3]

The demands of the iron industry and its ancillary services resulted in an intensive immigration of labour, initially from the surrounding rural areas and especially west Wales, but as time went on from England, Ireland and eventually even continental Europe. Some of these immigrants came to settle on a permanent basis, others were seasonal migrants who alternated work in the 'black kingdom' with work on the farms from which they originated. Living conditions in the iron towns may have been unpleasant and the work arduous and dangerous, but compared to agricultural labour the rewards were high (in the good times, at least) and town life offered an excitement and freedom that could never have been experienced in the closed communities that the villages represented.

In many cases the growth of the iron settlements was haphazard and uncontrolled, but there is also evidence of a planned approach to urban development at Tredegar or on a smaller scale at Butetown near Rhymney. There was a strong paternalistic element in the management of some settlements: Dowlais, for instance, was controlled by the Guest family who were earnest Methodists and Ebbw Vale, perhaps the most extreme example, developed under Quaker management into a company town with housing and all other facilities provided by the company. In other places company houses were provided only for the skilled artisans whom it was essential to attract and retain: a good example of such housing can still be seen at Blaenavon. Truck shops were operated by several companies including Dowlais, Tredegar, Ebbw Vale and Maesteg: these often came to be resented, but it could also be argued that in a newly created and often isolated community the company was the only organisation capable of sourcing and distributing food and other household goods on a sufficiently large scale. However, it did not take long for the settlements around the ironworks to develop into functioning towns. The retailers of essential basic commodities were the first to arrive, followed by the suppliers of a variety of more specialised services. The professions such as lawyers and bankers were

---

[3] For an account of the historical development of the iron industry, including statistics, see Ince 1993, Atkinson and Baber 1987, Rowson and Wright 2001. Rowson and Wright include an outstanding array of illustrations

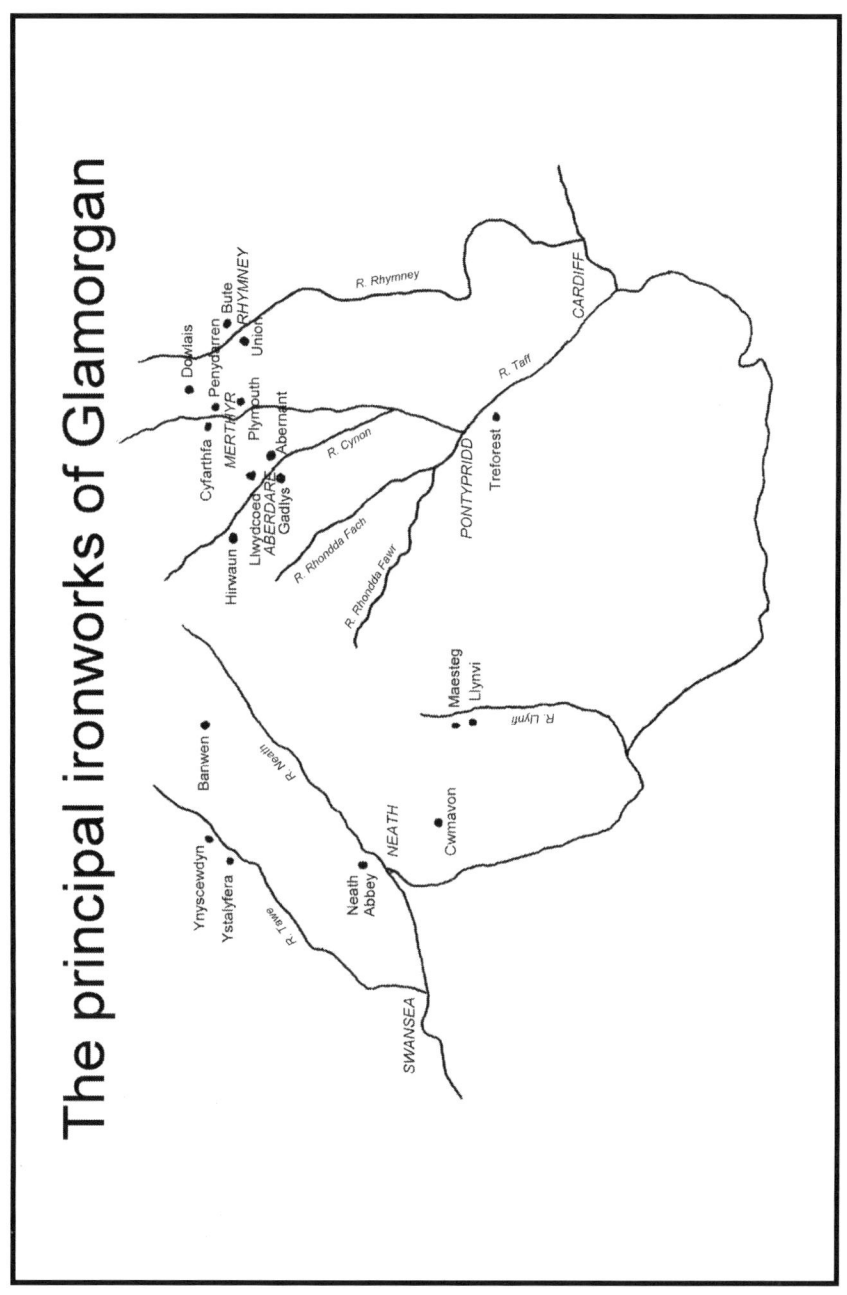

The principal ironworks of Glamorgan

slow to appear and even as late as 1850 could only be found in a few towns. Schools, newspapers and printers proliferated from early in the nineteenth century. Local administration slowly adapted to the new conditions. By the middle of the century the frontier settlements had developed into stable towns with the usual range of facilities, but in every case still dominated physically and economically by the local ironworks or collieries.

There was also a smaller scale iron industry on the southern outcrop of the coalfield, notably around Tondu to the north of Bridgend, and a rather less successful extension into the anthracite coalfield of western Glamorgan and Carmarthenshire in the nineteenth century. The only centre of iron production actually within the coalfield was at Maesteg where local geological conditions brought the coal and iron close to the surface.

The iron industry flourished for about a hundred years from the mid-eighteenth to the mid-nineteenth century. In the 1820s and 1830s it was responsible for around 40 *per cent* of the total output of pig iron in the United Kingdom. This proportion then began to fall even though in absolute terms output was increasing. The industry started to fail from about 1860: by then output represented less than 20 *per cent* of the total U.K. output and was declining, and long established ironworks were starting to be closed. By the 1880s the industry was in terminal decline although some works struggled on into the twentieth century. Local ironstone had become ever more expensive to extract since the easiest reserves had long since been worked out and consequently the south Wales industry was becoming increasingly dependent on imported ore. It was unable to compete with the newly developed iron industry centred on Middlesbrough with its much lower costs. At the same time the introduction of the Bessemer process made it possible to manufacture steel cheaply and in large quantities. This resulted in steel rapidly replacing the wrought iron which had always been the staple product of south Wales. Unfortunately the highly phosphoric ores to be found in south Wales were unsuitable for conversion into steel and the industry could only hope to survive by importing iron ore from outside. Transporting iron ore up the valleys made no economic sense and led to the iron industry of the *blaenau* being replaced by the steel industry of the coastal strip. The conditions that had led to the location of the industry on the *blaenau* in the first place were no longer applicable.

However, as the iron industry started to decline coal mining increased in importance. Coal had been worked within the valleys since at least the thirteenth century, and doubtless earlier than that, but as a high bulk/low value commodity overland transport to the nearest harbour was not economic and consumption remained local. The situation was rather different in the western part of the coalfield where the coal measures come right down to the coast. Here it was only a short distance from the pit to navigable water and so by the sixteenth century there was a flourishing export trade at ports such as Swansea, Neath and Llanelli, to say nothing of Pembrokeshire. Much of this trade was in anthracite for which a specialised market existed in all parts of the United Kingdom and not just in the Bristol Channel and Irish Sea zones.

With the establishment of the coke-fuelled iron industry along the *blaenau* there was obviously a vast growth in the output of coal but at first this was simply for conversion into coke to feed the furnaces. However, the opening of the Glamorganshire Canal in 1794 and even more so, of the Monmouthshire Canal in 1799, resulted in an increase in the output of coal as a sale commodity. Indeed, by 1830 the tonnage of coal carried on the Glamorganshire Canal was about 30 *per cent* higher than that of iron. Coal production was undertaken both by the iron companies as a profitable side line and by independent coal owners who had no connection with the iron industry. Technological developments also encouraged the growth of the coal trade. From 1828 the application of hot blast to the iron furnaces led to an increase in their efficiency and consequently to a reduction in the amount of coke that was required, thus releasing more coal for sale. Also, improved methods of colliery ventilation allowed pits to be sunk to a greater depth to exploit seams that had hitherto remained unworked. But perhaps the greatest factor in the growth of the south Wales coal industry was the development of steam navigation. The dry steam coals with their high carbon content that were to be found throughout the valleys were ideal for use in marine engines and as the navies and merchant marines of the world converted from sail to steam during the nineteenth century, so the demand for south Wales coal expanded.

Other factors that helped to encourage the development of the industry were the general adoption of gas for lighting and heating and the spread of railways which both required coal for the locomotives and made the distribution of coal for industrial and domestic use quicker and cheaper.

The result of all this was the development of deep mines and an increasing dependence on coal by the local economy. The Monmouthshire valleys had been exporting significant quantities of coal from about 1800, greatly helped by Newport's exemption from duty on exported coal, but this was largely from small-scale levels or shallow pits: the first deep mines were at Abercarn (1836) and Ebbw Vale (1841). At the same time, in the 1840s, there is the first clear evidence of coal starting to replace iron as the basis of the economy with the development of the very successful deep mines of the Cynon valley in place of an iron industry that had never been particularly successful. In the lower Rhondda valley Walter Coffin opened his first levels at Dinas in 1811. Development had reached as far up the valley as Tonypandy by 1845, but it was only in the 1850s that coal working throughout the two Rhondda valleys really developed. The coal industry continued to expand as the iron industry declined: new companies were set up and many of the old iron companies converted their operations into coal mining. The demand for steam coal ensured almost continuous growth in the south Wales industry, accompanied by population growth and urban expansion, until the peak year of 1913. Thereafter rapid decline set in, the social and economic consequences of which have still not been completely overcome.

~~~~~~

The history of the valleys has been extensively investigated by both academic and amateur historians but little attention has been given to the development of the postal service and its role in the successful operation of the iron and coal companies. This is surprising, given the way in which the iron and coal industries were organised. Typically iron (generally in the form of cast-iron or of wrought-iron bars) was produced at some point within the valleys. However, south Wales failed to develop an internal manufacturing industry (apart from tinplate in the western part of the coalfield and, from about 1830, railway rails). Instead, the product was sold to merchants in such centres as London, Liverpool or Bristol, or through agents located in major commercial centres throughout the British Isles. The iron was transported by canal or tramroad to the local port (in most cases Cardiff or Newport) where each company maintained its own wharf and where the shipping arrangements were made. by an agent based in the port. Similarly coal was sold through agents in the principal shipping ports such as Cardiff, Newport or Swansea.

A further factor is that in many cases control of a company was exercised from outside the region even though a member of one of the controlling dynasties might be resident locally. In the early nineteenth century Penydarren was owned by William Thompson, an important figure in City of London financial circles, and by William Forman whose family firm had manufactured ordnance in London for several generations. Slightly earlier Cyfarthfa had been owned by Richard Crawshay who controlled it from London until he moved to Merthyr in 1792 to take control in person. His grandson, William Crawshay, lived in style in Cyfarthfa Castle, but with his dictatorial father, also named William and who remained in London, always looking over his shoulder. Until their bankruptcy in 1842 the Quaker banking house of Harford & Co owned Ebbw Vale and controlled it from their base in Bristol. However, there were also owners who lived locally, such as Josiah John Guest at Dowlais, Samuel Homfray at Tredegar or the brothers Joseph and Crawshay Bailey at Nantyglo (where they felt it necessary to build themselves two fortified towers in which to take refuge in the event of serious unrest among the work force).

For an industry that was structured in this way to function effectively, it is clear that frequent communication between its different components was essential. Proprietors in London needed to maintain contact with their local managers to ensure that the business was being run efficiently and that their investments were being protected. Local managers need to report back on a regular basis and to seek guidance, not simply on strategic matters but often on very minor everyday issues: in the 1780s Richard Crawshay wrote frequent hectoring letters from London to his partner, James Cockshutt, who was manager of the Cyfarthfa works in Merthyr. Slightly later, William Taitt, a notorious worrier and micro-manager who controlled the affairs of the company from his base in Cardiff from 1787 to 1815, wrote several times a week to the local managers at Dowlais. Agents and merchants needed to send orders to the furnaces or to the pithead and the furnaces needed to be able to report on the progress of orders. Arrangements for despatching orders had to be made with the company's wharf or with the shipping agents. Complaints and enquiries had to be discussed and resolved; invoices, payments and receipts had to be processed. Legal business needed to be transacted, be it major events such as property leases or minor squabbles, and since the solicitors were in the older established towns on the periphery of the region such as Brecon or Abergavenny, routine business often had to conducted at a distance and not in person. If the company operated a truck shop there were supplies to be procured and this entailed communication with farmers or wholesalers.

Apart from the industrial concerns there were also the needs of those who lived in the communities around the works. There was the desire of the ironmaster's family to maintain contact with friends and family in other parts of the country, since in almost every case they were immigrants. As time went on and the ironmasters became involved in county affairs there was political business to be discussed and social arrangements to be made. Then there were the tradesmen and shop-keepers who had similar needs, if on a smaller scale, to those of the ironmasters: they needed to source their wares, to issue bills, to chase up defaulters, perhaps to manage legal affairs. There were the local administrators, the parish vestries: they may have been inadequate and frequently at a loss when it came to dealing with all the issues that industrialisation and population growth had thrown up, but they still needed to communicate, often with other vestries to make arrangements for sending paupers back to their home parish lest they became a charge against the rates. Least known, and perhaps the most interesting case of all, there were the ordinary people, the *gwerin*. Conventional wisdom has it that they never wrote letters – indeed, that they *could not* write letters because they were illiterate – but how true is this?

There was thus clearly a great need for a well organised means of communication both within the valleys and between the valleys and all other parts of the United Kingdom. This is illustrated by the extensive collections of correspondence which still survive in various archives, notably those originating from Dowlais, Cyfarthfa and Ebbw Vale. In the absence of telegraph, telephone, e-mail and all the other modern means of communication which are now taken for granted, the only means of communication at a distance was by the physical transmission of a written message, and before the railways the fastest speed at which this could be done was that of a galloping horse, about 10 m.p.h.[4] Industrial and commercial efficiency was thus highly dependent on some form of postal service, be it the official post operated by the General Post Office or a private post arranged to meet a particular local need, and it had to be conducted with the minimum of delay. For the major commercial user, cost was an important factor, but timeliness was of the essence.

The purpose of this study, therefore, is to examine the development of the postal service within the valleys of south Wales within the context of their industrial development. It attempts to show how the service developed to meet the demands that were placed upon it, especially by the ironmasters, since it was their activities that drove the local economy and it was their requirements that were of the uppermost importance. The period that it covers is approximately from 1760 to 1860. This period has been chosen on the grounds of factors related to both industrial activity and the postal service. Hirwaun ironworks was set up in 1757 and this represented the start of the coke-fired iron industry of south Wales. By 1860 the character of the south Wales industrial base was clearly changing: iron was in decline and the steam coal industry was expanding rapidly. Also by 1860 the effect of the postal reforms that were implemented from 1840 onwards and the widespread development of the railway network had led to the existence of what is recognisably a modern postal service. By that it is meant that overnight transmission of letters had become the norm (except in the case of long or involved journeys), with collections made in the late afternoon or evening and with deliveries the following day during the morning. Post offices existed in most communities of any size and letters were delivered free to most addresses (although it was not until 1900 that universal free delivery was achieved, regardless of the remoteness of a dwelling). Thus the period between 1760 and 1860 saw the development of the postal service in response to new patterns of industrial activity and of settlement, new forms of transport, and a new understanding of its role on the part of the General Post Office. By

---

[4] There was of course the optical telegraph, which was considerably faster, but examples were few in number and for dedicated purposes only

1860 these influences had all had their effect on the service and until the more recent changes brought about by mechanisation and automation and by the general introduction of modern communications technology, the operations of the Post Office remained stable. Changes and developments there were, but they were less dramatic and represented the fine tuning and enhancement of an established system.

This study concentrates on the way in which the Post Office tried to meet the requirements of its users, always subject to the constrictions of government policy from time to the time. The emphasis is on the provision of the infrastructure – the means by which the mails were transported, the routes which were served, the times at which the posts were despatched or arrived – since these were the factors that were of most concern to contemporary users. So far as the sources allow I have also tried to rescue from obscurity the individuals who operated the service. Their task was not always an easy one. One should spare a thought for the anonymous Merthyr postman of the 1760s who had to walk all the way to Brecon and back to collect the few letters for his village, or for William Dyke who did a circuitous walk of 27 miles from Pontypridd to Aberdare and back every day. On a sunny spring day it might have been pleasant enough for either of them, but hardly so in the rain and snow of winter. Or what about the driver of the mail coach from Cardiff that rolled into Merthyr at 10.00 p.m.? No doubt an enjoyable drive on a summer evening, but in the winter, when it was dark all the way, and the only light was the moon (perhaps) and the roads were rough and often icy it must have been not just unpleasant but positively alarming. We see other human foibles as well, such as Robert Jones, the Receiver at Aberdare, who cheerfully sorted the letters on the bar of his inn until he was found out by a horrified Surveyor, or Thomas Brockett, the Blackwood messenger who celebrated the marriage of the postmaster's daughter so enthusiastically that he lost his bag of letters. Then there was the eccentric Shôn Gwaun Adda of the Rhondda whose humorous wisecracks apparently once delighted his hearers but now seem utterly flat to us. Characters such as these who made the service work are at the heart of postal history. Without them there would be no historic letters for collectors to acquire or postmarks for them to examine.

Chapter 1

The overall structure of the Post Office

In 1760 there was no official postal service of any description within the valleys of south Wales. This is hardly surprising. The population was sparse and mainly dependent on subsistence agriculture. Much of the land formed part of large estates, but there were few gentry houses actually located within the region. The only place that could perhaps be called a town was Pontypool and that was on the very eastern edge of the coalfield. The only other settlement of any significance was Merthyr but that was no more than a large village. There were ancient parish churches but the small communities that clustered around them were no more than hamlets consisting of but a few houses – as is still the case with many of them. In other words there was an almost complete absence of what might be called the letter-writing classes: the solicitors were in the established towns that lay all around the valleys, such as Brecon, Abergavenny, Cardiff, Swansea or Newport. The clergy were for the most part unbeneficed curates standing in for absentee incumbents, badly paid and with a minimum of education. Estate administration was mostly carried out from without the region. Local administration, carried out by parish vestries, was just that, local.

There was little conception of the Post Office as the provider of a public service. The establishment of an official post had to be justified on strictly commercial grounds and if a post could not cover its costs and yield a profit it was simply not supplied. There was no question of cross-subsidisation of unprofitable routes from the surplus on other posts. That being so, there were only two official posts in south Wales in the middle of the eighteenth century. Both were operated by postboys travelling on horseback and both operated three times a week. To the north of the valleys was the post from London to Pembroke which ran via Hereford, Hay-on-Wye, Brecon and Llandovery. It connected at Gloucester with the region's other official post to Monmouth, Usk, Cardiff and Swansea. A more detailed examination of the postal routes into south Wales before and during the period covered by this book will be provided in chapter 2.

At this period the General Post Office was ultimately under the control of the Treasury, but directed by two joint Postmasters General, political appointees who mostly sat in the House of Lords. The Postmasters General had a staff of civil servants, the most senior of whom was the Secretary. This post increased in importance with the passage of time. Initially no more than a clerk, by the middle of the eighteenth century the Secretary had become the real director of the Post Office. By 1760 the administrative supremacy of the office was tacitly acknowledged and after 1765, instead of the Secretary waiting on the Postmasters General, he usually summoned them to the Board to accept his suggestions or ratify his decisions.[1]

Of the Secretaries to the Post Office the most influential was Sir Francis Freeling, the last to administer the old system before the reforms that culminated in the universal penny postage of 1840. He held the office from 1797 to 1836. In a debate in the House of Lords in 1836 the Duke of Wellington claimed that under Freeling's management the Post Office had been better administered than any post office in any country. He was described as possessing 'a clear and vigorous understanding ... and the power of expressing his thoughts and opinions, both verbally and in writing, with force and precision'.[2] But at the same time it has also been pointed out that Freeling's strength was as the administrator of an existing system. He was no innovator and his later years in office were clouded by an inability to understand the growing discontent with the traditional arrangements.[3] This can clearly be seen in Freeling's approach to the private post that operated from Newport to Tredegar and Ebbw Vale right up to 1839. It was run by the ironmasters Samuel Homfray, father and son, and was virtually a form of truck. It was an obvious anachronism, but despite frequent local representations concerning its excessive charges Freeling rejected all requests to introduce an official post, almost, it would seem, out of deference to the ironmasters and from a fear of offending their vested interests.

---

[1] Ellis 1958, 25
[2] Smith 2004
[3] Willcocks 1975, 70-1, 117

Freeling was succeeded as Secretary by Colonel W.L. Maberly, a former army officer who had been appointed specifically to implement improvements to the Post Office. However, his understanding of postal reform seems to have been decidedly Fabian, and he is perhaps best remembered for the enormous antipathy that he felt for Rowland Hill and his proposals. Both Freeling and Maberly appear frequently in the following chapters.

The Surveyors

Among the staff of the Post Office were six Travelling Surveyors who answered to the Resident Surveyor and so to the Board. These surveyors, first appointed in 1715, were each assigned a geographical area within which they were responsible for the maintenance of the service. They served as a link between the Board, the local function-aries and the public. They ensured that local postmasters carried out their duties efficiently and that they maintained their accounts accurately. If an individual postmaster neglected his duties, or carried them out in an unsatisfactory manner, the Surveyor was there to admonish and advise him. If he fell into arrears with his accounts it was for the Surveyor to issue an official warning and, if he proved recalcitrant, to arrange for his dismissal and, if appropriate, for his prosecution. The Surveyors were also expected to serve the government of the day in an intelligence gathering capacity by reporting on the local incidence of crime and disorder, economic conditions, elections, and so on.[4] They were also responsible for investigating complaints by members of the public in connection with alleged loss or delay to mail or over-charging. When requests for alterations or improvements to existing services were made by the public, or when new services were under consideration, the surveyors examined the feasibility of the proposal, concentrating especially on the costs and likely income, suggested the best way in which they could be implemented and reported their findings back to the Secretary who was generally content to advise the Postmasters General to accept their recommendations.

South Wales was the responsibility of the Surveyor of the Western District which also included the west of England and the west Midlands. It was a large area to cover and the Surveyor was constantly

---

[4] Ellis 1958, 61

on the move. For a young man the absence of immediate supervision and the travelling could have made it an attractive post, but as he grew older he must have found it increasingly burdensome. Nevertheless the job turnover was far from high. Samuel Woodcock was appointed in 1785 and he remained in post until 1823 when he was pensioned off. He was described as 'Seventy-seven years of age; thirty-seven years in the service; incapacitated by bodily infirmity'. He was replaced by Charles Rideout, then a clerk in the London Foreign Letter Office. Rideout remained in post for well over thirty years, but by 1860 he too was in receipt of the generous, but no doubt well deserved pension of £641 p.a. Although not directly connected with his duties in the western district, he is particularly remembered for having promoted the stamping machine patented by his son-in-law. The so-called Rideout machine was used in London in 1858-9 and again in 1866-7, but it was ultimately rejected by the Post Office in favour of the Pearson Hill parallel motion machine.

Postmasters/Deputies

One of the main responsibilities of the Surveyor was to ensure that the local postmasters, or Deputies as they were known in the eighteenth century, were kept up to the mark. A Deputy had to be 'respectable, literate and competent' and preferably to have his own business for reasons of economy. Many were innkeepers whose stables provided horses and postboys. Socially they ranged from mayors and aldermen to humble rural tenants. About one in five were women. They took an oath and had to give a financial bond. They received fixed allowances for the carriage of mail and salaries roughly in proportion to the level of business passing through the office.[5]

The kind of issues that the Surveyors had to deal with can be illustrated from Merthyr Tydfil: in 1841 the postmaster, a draper, had to be admonished for sorting letters on his shop counter and was instructed to provide a separate office for Post Office business 'forthwith'. In 1846, in an incident involving the same postmaster, the Surveyor was instructed to take charge of the office in consequence of the Deputy's arrears. Fortunately it did not come to that, for a week later it was

---

[5] Ellis 1958, 32

reported that the Deputy had made good his arrears: there was no question of misappropriation, it was simply neglect of duties.

Until 1854 Deputies were appointed by the Patronage Secretary of the Treasury on the advice of a person of some local influence.[6] This can be seen clearly, again in the case of Merthyr. When an official post was established in 1804 the appointment of the first postmaster was decided by the ironmasters, and in 1851, even during his last illness, Sir John Guest dictated a letter to his wife with his views on the appointment of a new postmaster for the town. Public opinion, however, also had its part to play: in 1847 Gwenllian Davies was appointed postmistress at Merthyr in succession to her late husband following a petition signed by all the tradesmen of the town.

Receiving houses (or sub-post offices) became more widespread following the postal reforms of 1840 and the creation of new settlements in response to industrial development. The old system of appointment still existed but it could not always be applied in the new valley communities. A note that often appears in the official minutes from the 1840s onwards is 'Surveyor to appoint' or similar, which means that a vacancy had been advertised, no suitable applicants had been forthcoming and so it was up to the Surveyor to find somebody who could be persuaded to take on the duties in question. In many offices there was a high turnover of sub-receivers and attempting to find someone willing and suitable to fill these vacancies must have been one of the more irksome of the Surveyor's tasks. In Dowlais there were no fewer than seven sub-receivers between 1843 and 1850, and there were other places where sub-receivers only lasted two or three years. Brynmawr had three between 1839 and 1845, and Pontypridd had five between 1830 and 1842 until the successful appointment of Charles Bassett who remained in office for over 25 years.

The reason for this high turnover is not absolutely clear, but it may well be that the sub-receivers found that the anticipated rewards were not commensurate with the commitment that was required. At Brynmawr the receiver had a salary of £10 p.a. and had to deliver about 440 letters a week as well as conducting the counter business: he asked for an increase in 1846 but was turned down. When a receiving house was set

---

[6] Daunton 1985, 276

up in Ebbw Vale in 1848 the Surveyor found it impossible to find anyone who was willing to undertake similar duties for the annual salary of £10. Where a separate letter-carrier was appointed and the receiver was not required to undertake the delivery, the receiver's salary might only be between £3 and £5 a year. There seems, too, to have been an element of backbiting among the local population which cannot have made the office of sub-receiver any more attractive. Complaints investigated by Rideout in 1840 turned out to be 'a malicious attempt to bring unfounded charges against the Sub Postmaster at Brynmawr'. A complaint against the receiver at Dowlais was made in 1844, but since it appeared that the signature on the complaint was fictitious no further action was taken. This suggests that the post of sub-receiver was not highly regarded and that appointments were sometimes made of necessity which turned out to be unsuitable. However, allegations of malpractice were not always unfounded. In 1845 the receiver at Dowlais was dismissed on the grounds that the Surveyor could not longer place any confidence in him 'after what has occurred'. (Details of the incident were not minuted, but perhaps the informant of 1844 had been onto something after all.) Shortly after that the receiver at Rhymney was in trouble for opening and delaying letters, and the receiver at Aberdare was warned not to sort the letters on the bar in his public house. In 1846 the sub-deputy at Tredegar had to be severely reprimanded for irregularities in the transmission of public money.

There was a similar high turnover of letter-carriers. Some were dismissed for incompetence (including one who was illiterate, not that the postmaster of Merthyr thought that that mattered), others resigned because they felt that their wages of 12s or so per week were not sufficient reward for the long hours and lengthy walks in all sorts of weather. In these cases, too, it often fell to the Surveyor to make an appointment.

Local arrangements

Local delivery arrangements could take a variety of forms. Legally the Post Office had an obligation to provide a free delivery to all houses within the limits of every post town, but in practice this was rarely provided until the nineteenth century. As might be expected, Merthyr was the first place in the valleys to have an official free delivery which started in 1837 and replaced an earlier private delivery service carried out by the postmaster for which he charged ½d per letter. As part of the programme of postal reform it was decided in 1843 that 'wherever a village or hamlet, or walk through which the postman should go, should have 100 letters addressed to it, that district should be entitled to a post at the public expense'. The result was a great increase in the number of sub offices and free deliveries and by 1859 it was claimed that about 93 *per cent* of letters were being delivered free.[7] A general revision of these rural posts commenced in 1851. Anthony Trollope, who as is well known earned his living in the Post Office as well as being a prolific author, was responsible for carrying out this task in south Wales. He started work in 1852 when his request for a set of Ordnance Survey maps was approved and he completed it in 1855. His reports are preserved in the Post Office Archive.[8] Under each post town he lists each of its rural posts, their frequency, the places at which deliveries were made, and the remuneration of the letter-carriers and the sub-receivers.

For the unreformed Post Office it was essential that any post covered its costs and provided a surplus for the Revenue. Sometimes, as in the case of the first official post to Merthyr, a post might be set up on an experimental basis with a view to discontinuing it if it failed to produce the anticipated income. Where an official post could not be maintained profitably local communities might offer a guarantee to the Post Office against any losses: thus as late as 1841 Cwmtwrch in the Swansea valley was told that an official post could only be provided if the inhabitants guaranteed to meet the costs.

An alternative measure was a Fifth Clause post under which the Post Office would provide the service at an agreed charge to the local residents. This arrangement was not generally popular. There were no posts of this category in south Wales although the possibility of making the Merthyr post a Fifth Clause post in 1804 had been considered and rejected, apparently on Freeling's recommendation. When Pontypool asked for an official post in 1807, all that was offered was a Fifth Clause post which was turned down by the inhabitants.

---

[7] Daunton 1985, 43-4
[8] POST 14

Private posts

The strictly economic principles underlying the official post led to the creation of a great many private posts in all parts of the United Kingdom, the valleys not excepted. A full list of the private posts known to have operated in the region is given in Appendix 1. So far as the Post Office was concerned their responsibilities ended with delivery to the post office shown in the address. After that it was up to the addressee to make his own arrangements. This meant that an individual could choose the post office to which he wished his letters to be sent. In 1824, for example, following the introduction of a mail coach from Abergavenny to Merthyr, the Ebbw Vale Company requested their correspondents in London, Ireland and the north of England to address their letters to the company at Abergavenny rather than at Newport.[9] In the case of outgoing letters the Post Office would collect any mail handed in at an established post office, but how it got there in the first place was no concern of theirs. Again, individuals and communities had to make their own arrangements.

These arrangements could take a great variety of forms.[10] A messenger could be employed to carry letters to and from the local post town, and he might be paid by annual subscription or from fees on individual letters or by a combination of both methods. The messenger might be employed by the local postmaster, by a local resident or by the residents in general, or he might be a self-employed entrepreneur. In the iron district several instances are known of a private post being conducted by the iron company, more or less as part of the business. Examples include Tredegar, Blaenavon and Ebbw Vale. Mail coaches and stage coaches might carry letters from a post town to places along their route to which there was no official post as a private arrangement: since there was no question of diverting income from the Revenue this was a matter of indifference to the postal authorities. This was probably the arrangement adopted in 1824 for mail from Abergavenny to Ebbw Vale. Gentry families and large businesses would probably have their own private bag and arrange for it to be taken to and from the post by a family servant or a company employee: the postmaster would hold one

---

[9] Reynolds 2007
[10] Some of the arrangements under which private posts operated are described in a Treasury minute of 1841 reproduced in Willcocks 1975 149-52; cf Daunton 1985, 41

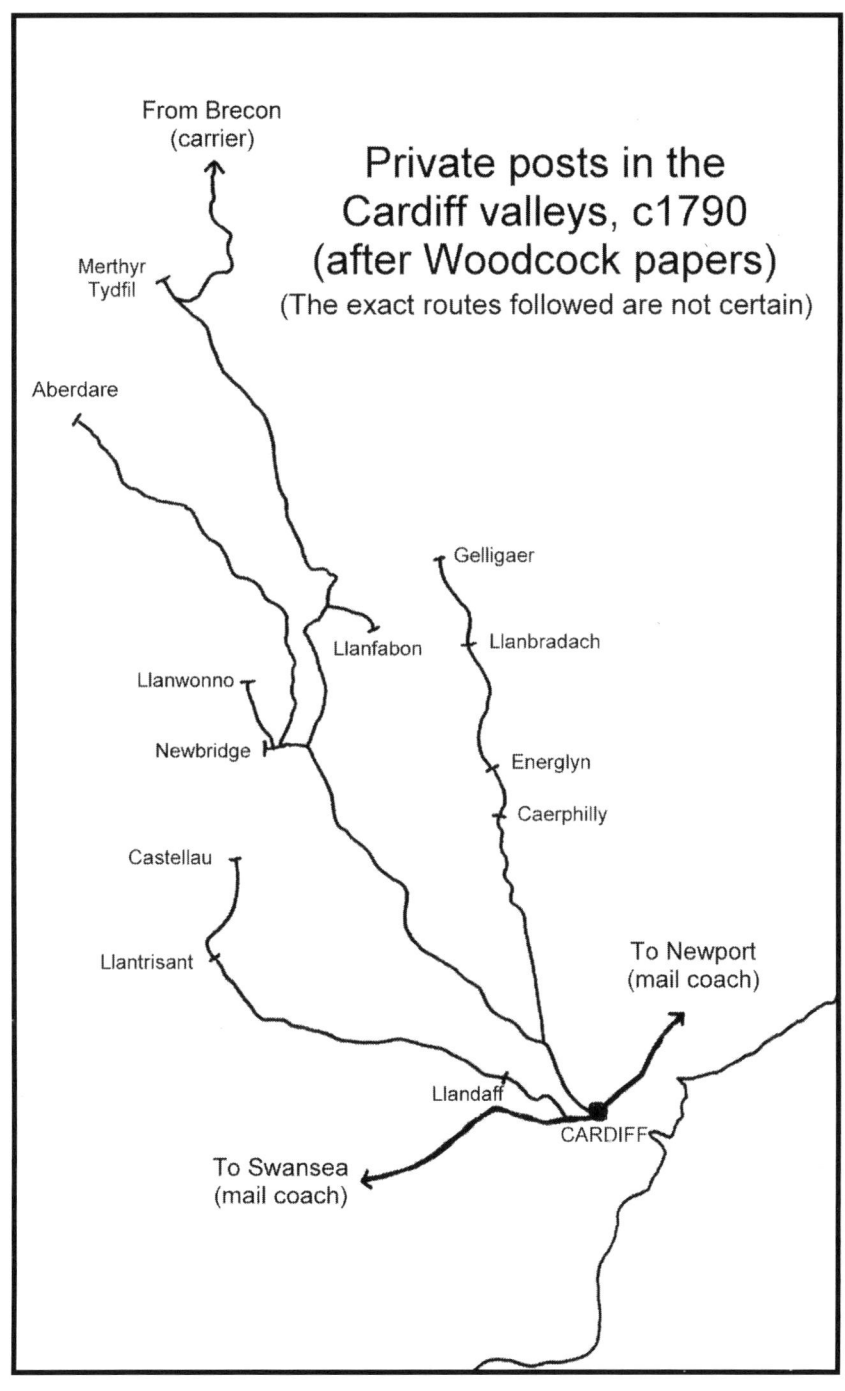

Private posts in the
Cardiff valleys, c1790
(after Woodcock papers)
(The exact routes followed are not certain)

key, the owner of the bag, the other. Most of the ironmasters of Merthyr had their own bag, but William Taitt declared, rather self-righteously, when the official post was started in 1804, that he was quite happy for Dowlais company mail to go in the 'common bag'. These arrangements suited heavy users of the post well enough, but the occasional user generally depended on a visit to the post office to collect or post his letters, either in person or through the agency of a friend or servant. Even those who normally used a private post might sometimes choose to send letters by friends, servants, or carriers if this was cheaper, quicker or more convenient. Indeed, this was sometimes the only choice. The ironmasters of Merthyr had frequent occasion to correspond with solicitors in Brecon, but the official post via Cardiff and Carmarthen was slow and it seems that after 1787 and until 1823 the only link with Brecon was a carrier who went there and back once a week. Consequently there was no choice but to use private messengers.

The private post contractors naturally charged a fee for their services. It might be as low as 1d per item or it could be as high as 6d. This was over and above the postage charged by the General Post. These additional charges were often resented and led to local demands for the establishment of either an official post or at least an official Penny Post. The hope was that if an official post were granted, the additional distance that an item had to be carried would be so small that the postage would still fall within the same charging band as to the post town and there would thus be no supplementary charge for local delivery. In the case of a Penny Post letters were collected and delivered within the stated catchment area for a flat fee of 1d per item over and above the General Post charge, but this could still be less than the charges raised by a private post. (For collection and delivery within the same area the fee was 2d.) Several Penny Posts were established in the valleys before 1840, including posts based on Merthyr, Abergavenny, Cardiff, Newport and Neath.[11] Local Penny Posts were superseded after 1840 by universal penny postage.

Of all the private posts that operated within the valleys the only one of which we have anything like adequate knowledge is that operated by the Homfrays between Newport, Tredegar and Ebbw Vale. It was in

---

[11] Archer 1987, 62-3

existence from the early 1800s until 1839 when the Post Office finally plucked up the courage to face down Homfray and institute an official post. We are indebted for this information to a few vociferous local residents who had little choice but to use Homfray's post and whose complaints have been preserved in the Post Office Archive. It is also possible, thanks to the Dowlais letters now preserved in the Glamorgan Record Office, to understand a little of way in which the private post between Cardiff and Merthyr operated in the period 1787-1804, and similarly the diary of Charles Wood[12] reveals the existence of an earlier post from Merthyr to Brecon. Documents in the Post Office Archive indicate the existence of a few more local posts, generally in connection with their replacement by an official post. There are also fragments of information preserved within the oral tradition and recorded by local historians of the nineteenth century such as Charles Wilkins and 'Morien' and by other printed sources. But most of the private posts are represented by no more than a name in the survey of all the post towns in his district which Samuel Woodcock started to compile in about 1790. Sadly, all of these accounts leave many questions unanswered as to the way in which the posts were operated and financed, what charges were levied and how they were collected, and what arrangements were made with postmasters and customers.

There was a tendency in later years to regard these private posts as quaint and inefficient. This is typified by the picture which Charles Wilkins gives of the old woman who brought the letters from Brecon to Merthyr and then spread them out on her kitchen table for the villagers to peruse.[13] Her operations may have been on a vastly smaller scale than those of the post office which Wilkins ran a hundred years later, but that is not to say that she was any less efficient, or that the Post Office of the day would have provided any better a service. There were, of course, messengers who were unreliable: an earlier postman at Merthyr was described by Charles Wood as 'a drinking man' with the implication that he could not be depended upon, but on the other hand there were devoted public servants such as William Dyke who walked from Pontypridd or Abercynon to Aberdare and back, 27 miles every day, for sixteen years and was still doing so in 1831 at the age of 62.

---

[12] Wood 2001
[13] Wilkins 1908, 499 (reproduced on p 89 of the present work)

And certainly some of the incidents that have been recorded concerning official letter-carriers show that the official post was just as liable to human failings as any private post.

To complement the private posts there were also private receiving houses. These might have been particular points, perhaps a specified inn or cottage, at which the postman would leave or collect letters. The Tredegar post had a string of private receiving houses including Risca, Pontllanfraith and Blackwood. In other cases the keeper of a private receiving house would meet a mail coach or an official postman to receive and hold letters for his locality. On the road from Cardiff to Merthyr there were several such receiving houses including Quakers Yard and Newbridge (now Pontypridd). The latter was suppressed in 1830 after the keeper had been found to have been tampering with the postage marks. There was also a private receiving house in Dowlais. When it was replaced by an official post office the keeper asked for compensation. His request was not successful but he was appointed official sub-receiver. Similarly in the Swansea valley many of the receiving houses set up by the operator of a private post were converted into official sub offices. There must have been many other private receiving houses which have left no trace in the historical record.

One of the duties of the Surveyors which they were increasingly called upon to perform was to examine requests from the public for the establishment of new posts or for improvements to existing posts. These requests might come from local landowners and gentry, from Members of Parliament, or from industrialists and other commercial users; they could also take the form of petitions presented by groups of tradesmen or by the inhabitants in general of a place. Petitioners were also very much concerned with existing services and urged measures that would reduce the time taken by a letter, that would increase the reliability of the service, or that would result in more convenient timings. Reducing the cost to the user was obviously desirable and a welcome side effect of any change, but improvements in terms of speed, reliability and times were the uppermost consideration, certainly for commercial and professional users.

The ability to receive letters quickly and promptly and to reply without delay are very understandable concerns in an environment in which the post was the only means of communication. 'By return of post' meant just that: the lawyer or industrialist wanted to be able to receive a letter by the incoming post and have time to digest its contents and reply by the next outgoing post. This was not always possible. The posts might arrive or depart at inconvenient times during the night; the time between arrival and departure might be very brief. This was simply a reflection of the fact that the timing of any one post depended on the whole network of inter-related services throughout the United Kingdom. Their timings in turn were influenced by uncontrollable factors such as weather conditions and topography, but above all by the demands of the City of London that outgoing mails should leave in the evening and incoming ones arrive in the early morning. Consequently working hours in the offices and counting houses in the rest of the country had to fit in with the times of the post, whatever inconvenience that might cause. Any change to the times at which mails arrived or were despatched and which resulted in either a greater length of time between arrival and departure or in a better match with normal working hours was always welcomed.

Because the times at which the post arrived and departed were a matter of such great concern to contemporary users, I have given close attention to this subject in the following chapters. Some readers may think that I have gone into excessive detail but I make no apology for this. A clear knowledge of the timings is necessary in order to understand how well the postal service met the requirements of its users. It also makes it possible to understand how the whole network of posts intermeshed to provide a service for any one place to and from all other parts of the country, and so explain why a local service was run to whatever its particular timings might have been.

Chapter 2

External circulation of mail to south Wales

In order to understand the circulation of mail within the south Wales valleys, it is necessary to understand how mail reached the region from other parts of the United Kingdom. Without this knowledge it is not possible to appreciate fully the reasons for the changes that were made to the local arrangements and the benefits which they were intended to bring. Similarly, unless one understands the parameters within which the Post Office had to work – basically the requirement to provide an interconnected national service at the minimum of cost – the reasons behind an apparently less than perfect decision may not be understood. This chapter, therefore, sets out the overarching framework within which local developments were made in south Wales up to about 1860. It covers the transition from horse-mounted postboys to mail coaches in the 1780s, the growth and expansion of the mail coaches, and then their decline from the 1840s as the carriage of mail started to be transferred to the public locomotive railways which were able to provide a service which was recognisably modern in its speed and timing.

An organised postal system was created in the 16th century exclusively for the carriage of royal mail. This included a post from London to Bristol. It was temporarily extended through south Wales to Haverford-west (via Cardiff and Swansea) in 1600 at a time when there were fears of Spanish military intervention in Ireland. When peace was made in 1603 there was no longer any need for a post to Bristol and south Wales and it was discontinued soon afterwards.[1] There is then no evidence for an official post in south Wales for another fifty years, although it may well be that some sort of private service existed, as happened elsewhere in the first decades of the 17th century.

---

[1] Beale 1998, 183 dates the abandonment of the Haverfordwest post to 1603 (p.210) but elsewhere quotes an annual cost for the service in 1605 (p.237). Kay 1951 dates the abandonment of this post to 'some twenty years later' from 1598

King Charles I opened the posts to the public in 1635. His motive was partly financial, but it also gave the authorities an opportunity to examine correspondence from suspected dissidents. The royal pro-clamation which entrusted Thomas Witherings with the organisation of the service specified a post from London to Bristol but no further.[2] In about 1649, in the aftermath of the Civil War, Sir Edmond Prideaux set up a postal service for the use of the state and its officials, which seems to have extended as far as Milford Haven.[3] Soon after, in 1653, the first mention appears of a postmaster in south Wales, Roger Bayley of Swansea. He was one of a number of postmasters who submitted a petition to the government for the purpose of safeguarding their position following the decision in that year to re-open the post to general use. Their petition must indicate that a regular post had been established between London and south Wales by 1653.

The decision of the Parliamentary regime in 1653 was endorsed by Charles II in 1660. The south Wales mail ran through Gloucester, Monmouth, Usk, Cardiff, Swansea and Carmarthen to Pembroke. However, as a result of the unreliability of some of the postmasters on this route, the Pembroke mail was re-routed through Brecon in 1674. This left Cardiff and Swansea unprovided for and to meet their requirements a by-post from Monmouth to Swansea was established.

The post ran twice a week, leaving London on Tuesday and Saturday evenings, with corresponding up posts due into London on Mondays and Fridays. The postboys who carried the letters were mounted on horses and they were expected to maintain a speed of 5 m.p.h. On this basis Swansea was at least 42 hours distant from London and Pembroke 55 hours. As often as not, however, these times probably represented an aspiration rather than a normal achievement. The post was still only twice-weekly in 1755 but by 1765 a third post had been added, so that the posts to south Wales from London now left on Tuesday, Thursday and Saturday. Also, by this date, the routes had been revised. The Pembroke mail ran from Gloucester to Hereford and then on to Brecon and Carmarthen; the Swansea by-post started at Gloucester.

---

[2] Facsimile reproduced in Beale 1998, 228-9
[3] Willcocks 1975, 13-14

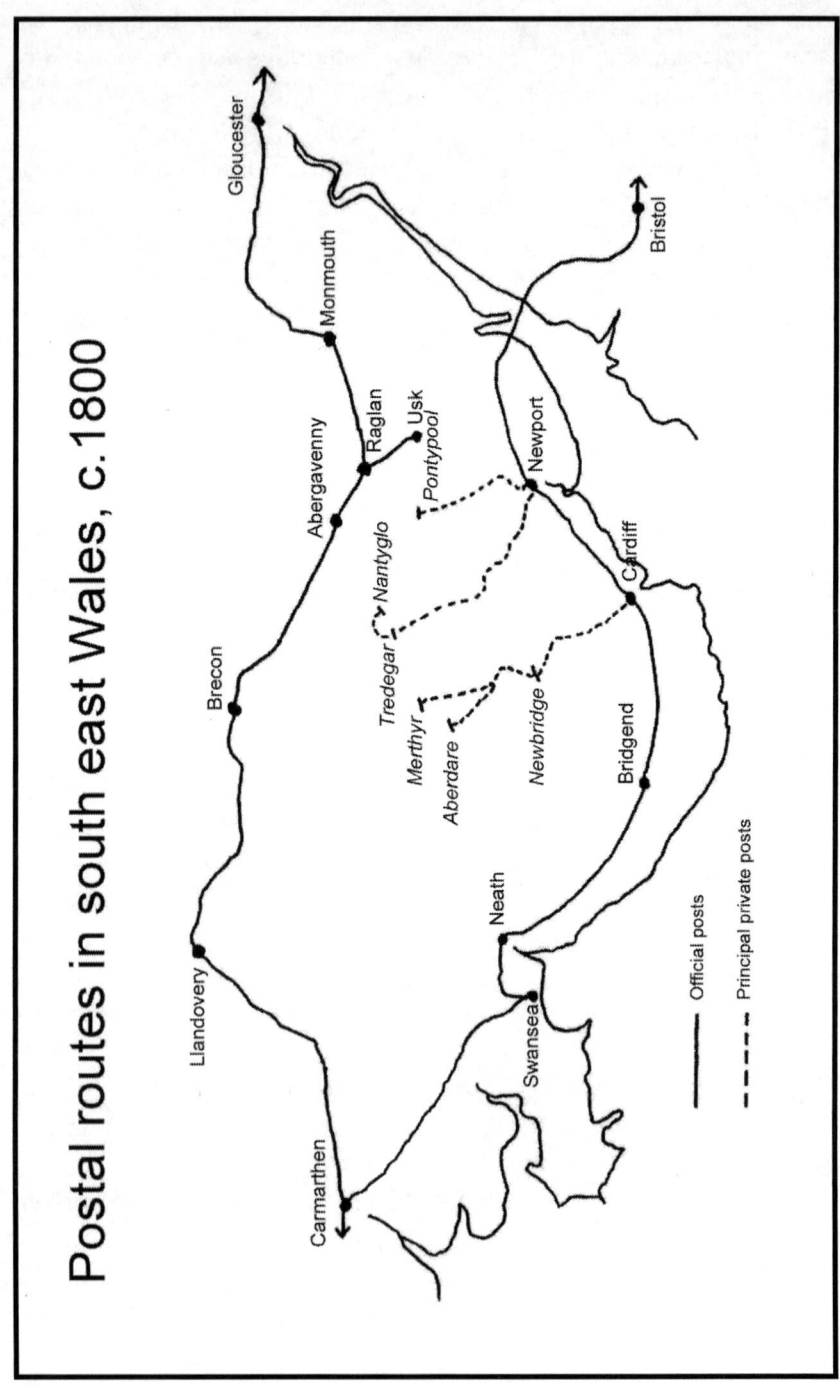

Postal routes in south east Wales, c.1800

Needless to say there was no official post of any kind within the south Wales valleys, but it is more than likely that a number of private posts operated for the convenience of communities and gentry houses that lay away from the mail road. There is clear evidence for such a post that linked Merthyr to Brecon once a week in the 1760s and by 1790 Woodcock was able to record a host of private posts under each post town in his district. In the 1760s, when the first iron furnaces started to be built at Merthyr, the time taken for a letter from London to Merthyr would have been nearly 48 hours. It would have taken at least 36 hours for the 180 miles to Brecon via Hereford, assuming that there had been no delays *en route*. Assuming too that the private postman from Merthyr had turned up on time and was ready to go back as soon as the letters had been sorted (allow an hour for that), it would have taken him about five hours to walk back across the Beacons, giving a total of 42 hours so long as there were no delays at any stage of the journey.

In 1784 a radical change was made to the way in which mail was circulated, with 'Mr. Palmer's plan' for using mail coaches on all the arterial routes from London and on the main cross-country routes. These provided a greatly improved service, in terms of both speed and reliability, compared with the horse-mounted postboys. A coach from London to Gloucester was started in August 1785. The following October it was extended through Hereford, Brecon and Carmarthen to Hubberston (the precursor of Milford Haven). At the same time another mail coach was started from Gloucester, in connection with the Hubberston mail, which worked its way through Chepstow and Cardiff to Swansea. This coach was almost immediately extended to Carmarthen to meet the mail on the northern route. Thus by the end of 1785 south Wales was connected to London and the rest of the kingdom by two mail coach services, London to Hubberston via Gloucester and Brecon, and Gloucester to Carmarthen. Both ran on a daily basis.[4]

In August 1787 the main service to Hubberston was re-routed via Bristol, Cardiff and Swansea as a result of the introduction of the official packet service between Milford Haven and Waterford. The road from Gloucester to Carmarthen through Brecon reverted to a horse

---

[4] Beaver 2005

post. The new route was shorter but it involved the dangerous crossing of the Severn by the New Passage between Redwick and Sudbrook, more or less on the line of the later Severn railway tunnel. In 1797, therefore, after repeated public protests, the service was re-cast once again. The direct service between Bristol and Swansea was too important to be abandoned completely, so by way of a compromise the mail coach now ran on alternate days on the northern route (via Gloucester and Brecon) and on the southern route (via Bristol and Swansea). On non-coach days there was a horse post on each route so as to maintain a daily service.

By 1804 the mail coach on the southern route had been upgraded to a daily frequency between Bristol and Swansea, although the continuation from Swansea to Carmarthen was still made by coach and by horse on alternate days. In 1805 the mail coach on the northern route was also made daily. There was thus a daily coach from Gloucester to Milford Haven on the northern route (the modern A40 road), and from Bristol to Swansea on the southern route (the modern A48 road), with an extension to Carmarthen three days a week. In 1807 or 1808 this was increased to a daily frequency. There were now daily coaches on both the northern and the southern roads; they met at Carmarthen where the mails were combined and carried on to Milford Haven on one coach. These arrangements continued with very little change until 1835. Arrival times were advanced as roads were improved, and the overall times were reduced. Until 1822 the journey from London to Swansea took 32 hours or more. In 1822 this was reduced to about 26 hours, and in 1837, even with further improvements, a letter still took about 16 hours to get from London to Cardiff and about 21 hours to get to Swansea – and that was if the mail coach kept to time, which very often it did not. It was claimed, for instance, that in the winter of 1834/35 the mail arrived on time in Swansea on only one day in five. It was frequently between two and five hours late and on occasion as much as seven hours late.[5] The coach on the northern route between Gloucester and Carmarthen was timed at an average speed of over 10

---

[5] *The Cambrian* 14 February, 28 March 1835

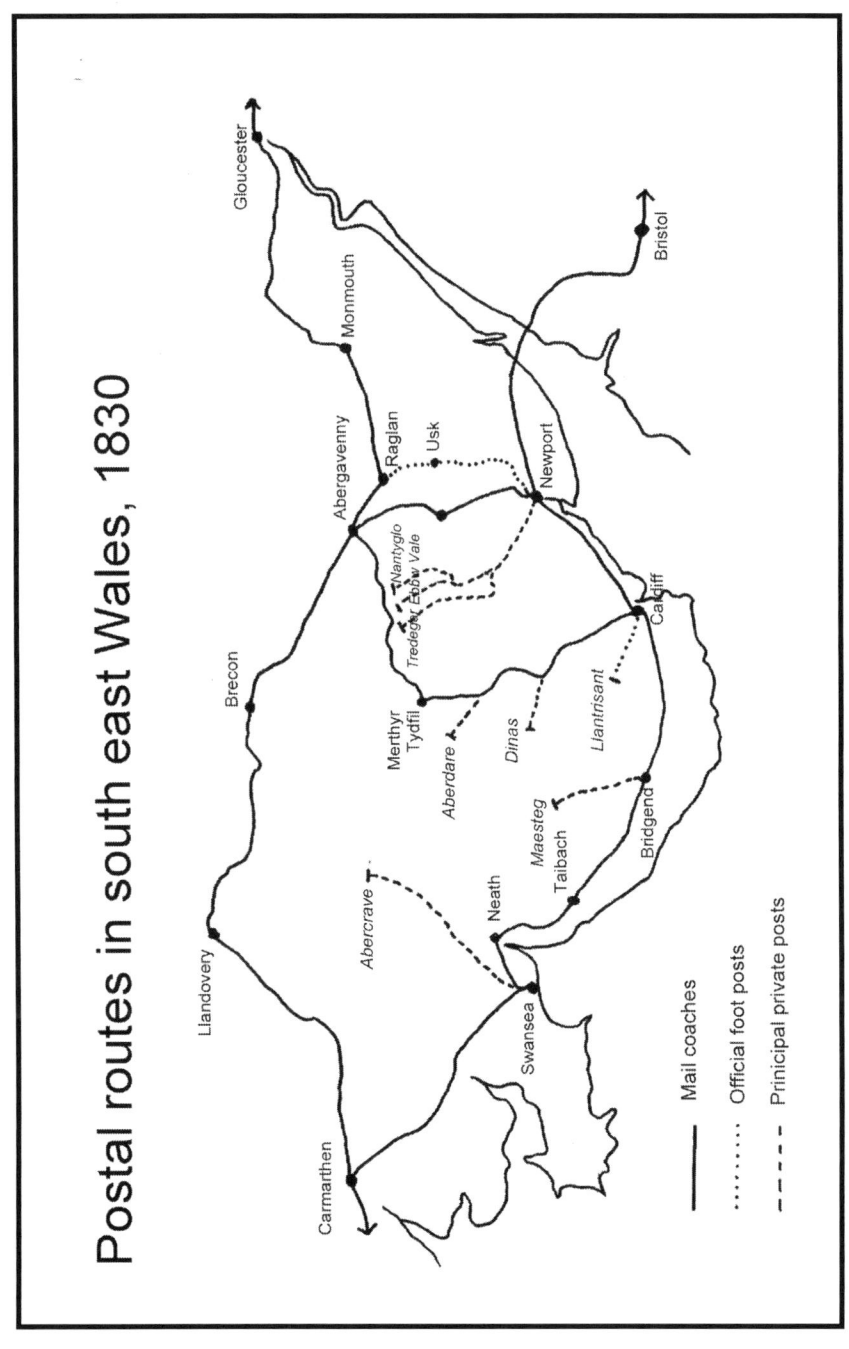

Postal routes in south east Wales, 1830

m.p.h. in 1838, while on the southern route the speed was exactly 10 m.p.h., which must have excluded the time taken to cross the Severn.[6]

During this period, too, official posts started to appear within the valleys, their timings designed to connect with the London coaches. The first of these was a horse post from Cardiff to Merthyr in 1804. It was converted to a mail coach in 1821. Mail coaches were also introduced between Abergavenny and Merthyr in 1824 and Newport and Abergavenny in 1827. Even so, by 1830 the only places within the valleys that enjoyed an official post were Merthyr, Pontypool and Newbridge (later to be renamed Pontypridd). There were, however, a number of well organised private posts.

One of the main causes of delay to the mails was the crossing of the Severn at the New Passage. The bags were taken across in an open boat which was at the mercy of the weather. There was no proper landing place and so passengers and the mail guard faced an unpleasant, not to say dangerous, experience, in getting into and out of the boat. According to John Henry Vivian, M.P. for Swansea, after ' ... three miles ... in an open boat [they had] to wade through the mud and water over slippery rocks'.[7] For many years Vivian put himself at the head of a campaign to improve the mail service between south Wales and London. As well as being an M.P. he was also a leading figure in the Swansea copper-smelting industry and owner of the Hafod copper-works, one of the largest concerns in what was then a highly successful industry. In 1833 he proposed that the coach from Abergavenny to Merthyr, which had been established in 1824 and gave Merthyr the benefit of a London mail that was not at the mercy of the Severn crossing, should be extended to Swansea. Committees were set up in Swansea, Neath and Merthyr to push forward the idea of an Abergavenny to Swansea mail coach, memorials were signed and meetings were held.[8] The Post Office turned the proposal down, partly on the grounds that the state of the roads in question made it impossible to run a mail coach, but mainly because it would affect the viability of the mail coaches on the two main roads. However, by way of a consolation prize they agreed that a more direct service should be set

---

[6] 'Returns relating to mail coaches, etc.'
[7] *The Cambrian* 8 February 1834
[8] *The Cambrian* 10, 24, 31 August, 14 September 1833

up between Swansea and Merthyr, which duly took place early in 1834 with a mail cart from Neath to Merthyr, upgraded in 1835 to a mail coach from Swansea.

Following continuing complaints in south Wales and continued lobbying by Vivian, the Postmaster General finally agreed to transfer the mail from the New Passage to the Old Passage which ran between Aust and Beachley, more or less on the line of the first Severn road bridge. The Old Passage had been much improved in recent years with new stone piers on both banks and with the introduction of steamboats, the first in 1827 and a second in 1832. This, it was believed, would provide an easier crossing of the Severn and so ensure a more reliable service. Opinion in Swansea, and no doubt elsewhere, would have preferred the London mail to have been routed via Gloucester and Chepstow, so as to avoid the Severn altogether, but the importance of maintaining a direct link with Bristol ruled out this solution. The new arrangements took effect from 5 April 1835 and reduced the time from London by nearly two hours.

However, the Old Passage proved to be no more satisfactory than the New. 'Government steamers' had been promised in 1835 to carry the mail, but that was an empty promise. Mail was still carried in open boats, either rowed by men or towed by a steamer. The mail coach contractors, not unreasonably, refused to cross the estuary in the dark under these conditions and so the mail from London that, since July 1841, had arrived at Bristol by train at 1.15 a.m. was being held for over four hours before it was sent on by road at 5.50 a.m. In any case, even if the contractors had been willing to cross the Passage in the dark, the mail would still have been held at Bristol until the Portsmouth mail had rolled in.

Vivian continued to lead the campaign for improvements. In 1841, at his prompting, the magistrates of Glamorgan presented a collective memorial to the Postmaster General requesting that the London mail be routed through Gloucester, with an 'auxiliary mail' to maintain a direct link between Bristol and south Wales. This was turned down in 1842 and Vivian then proposed that the piers at the Old Passage be lengthened so that steamers could be used whatever the state of the tide. If steamers were used the contractors could no longer object to making the crossing in the dark and a much earlier start from Bristol would be possible. A letter could be replied to by return of post – although in the case of Swansea only just, since under Vivian's proposals there would be but one hour between the arrival of the down coach and the departure of the up coach.[9]

Meanwhile the Great Western Railway was building its line to Gloucester. South Wales realised that this was an opportunity to obtain a real improvement to the London mail. Vivian continued to lobby the Post Office and during the winter of 1844/5 opinion throughout the district was mobilised to have the London mail transferred to the Gloucester route as soon as the railway was open.

At first the Post Office was far from responsive. They claimed that they had already advertised for a contractor to horse a coach from Gloucester to Carmarthen via Chepstow, but without success. Vivian pointed out that the only contractor they had approached had been Mr Niblett, a leading coach operator based in Bristol who could hardly have been expected to show any interest. Perhaps rather more persuasive was the objection that routing the London mail through Gloucester would create difficulties with the North mail. Since February 1841 the North mail had arrived in Gloucester from Birmingham by train at 3.40 a.m. It was then sent on by coach to Chepstow where it connected with the coach that had started from Bristol at 5.50 a.m. If the London mail were to be sent from Gloucester there would be three possible ways of handling the North mail, all equally unattractive. The London mail could be held at Gloucester for two hours which would negate much of the benefit of using the new route and avoiding the Passage; the North mail could be held until the following morning, i.e. delayed for 22 hours, and sent on with the London mail as soon as the train arrived; or a second mail could be put on for the North mail, which would add to the costs of the Post Office.

The line to Gloucester opened on 12 May 1845 and new arrangements for south Wales mail were introduced on 1 September. They were rather better than anyone might have expected, given the previous response of the Post Office. The London mail was sent to Gloucester by rail, arriving there at 1.45 a.m. It was then immediately forwarded by coach, arriving at Cardiff at 7.39 a.m. and at Swansea at 12.25 p.m., an

---

[9] *The Cambrian* 19 February 1842

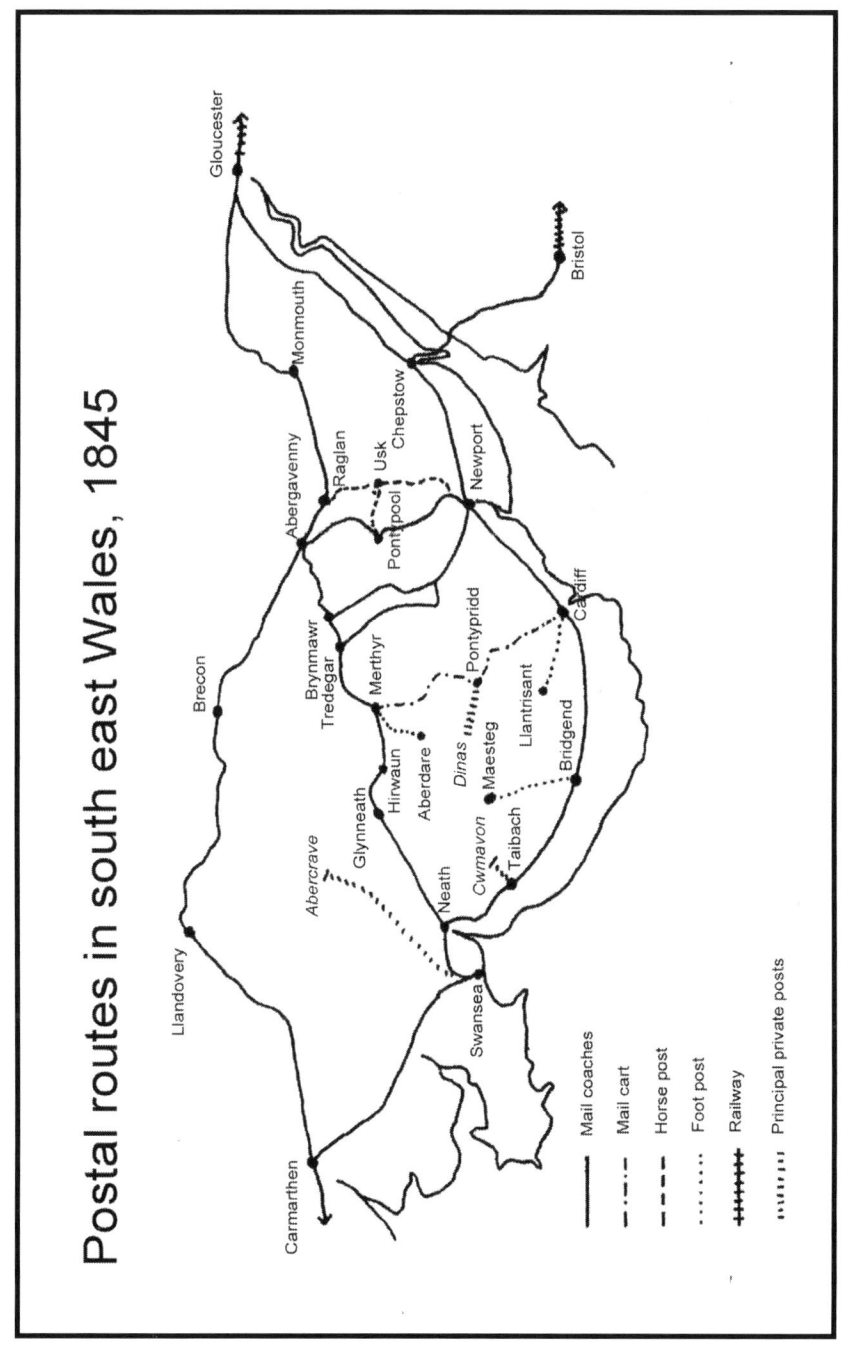

Postal routes in south east Wales, 1845

improvement of about five hours. The coach continued to Carmarthen and replaced the Bristol-Carmarthen coach. A mail cart left Bristol at midnight with mail from the west of England and crossed the Old Passage to Chepstow where it met the down mail from Gloucester at 4.45 a.m. The important link between Bristol and south Wales was thus maintained: the operator of the mail cart apparently had no qualms about crossing the river by night. The North mail was carried all the way to Bristol by train and was then sent on as a separate mail which followed the London mail through south Wales several hours later. An existing stage coach, Mr Niblett's *Cymro*, which ran as far as Swansea, was used for this purpose. It left Bristol at 8.00 a.m., crossed the Severn by the Old Passage and arrived in Cardiff at 1.42 p.m. and Swansea at 7.00 p.m.[10] Thus there were now two separate mails on the southern route across south Wales, the London mail followed several hours later by the North mail. This pattern was to remain essentially unchanged for a good hundred years.

Following the opening of the South Wales Railway from Chepstow to Swansea on 18 June 1850, the Post Office transferred the mails for places between Chepstow and Swansea to the railway as soon as possible. This took place on 6 July for the North mail and on 27 July for the London mail. The three weeks delay for the London mail was because of the need to give the contractors a clear month's notice. The North mail was now put off the train from Birmingham at Gloucester; it was sent on to Chepstow by road before transferring to a train which left Chepstow at 12.00 noon. The London mail reverted to the Bristol route and crossed the Severn at the Old Passage. It left Chepstow at 4.45 a.m. and reached Cardiff at 5.45 a.m. and Swansea at 7.45 a.m. The prospect of the London mail using the Old Passage was not welcomed, and indeed, by the end of August, even before the winter weather set in, serious delays had already arisen.[11]

However, this was only a temporary annoyance, since the railway from Gloucester to Chepstow was opened in September 1851 and the mails were transferred to this new route as soon as the existing contracts expired. The North mail was first: from 22 October it was transferred from the Birmingham-Bristol train at Gloucester and sent on by rail to

---

[10] *The Cambrian* 16 August 1845
[11] *The Cambrian* 30 August 1850; *The Times*, 9 September 1850

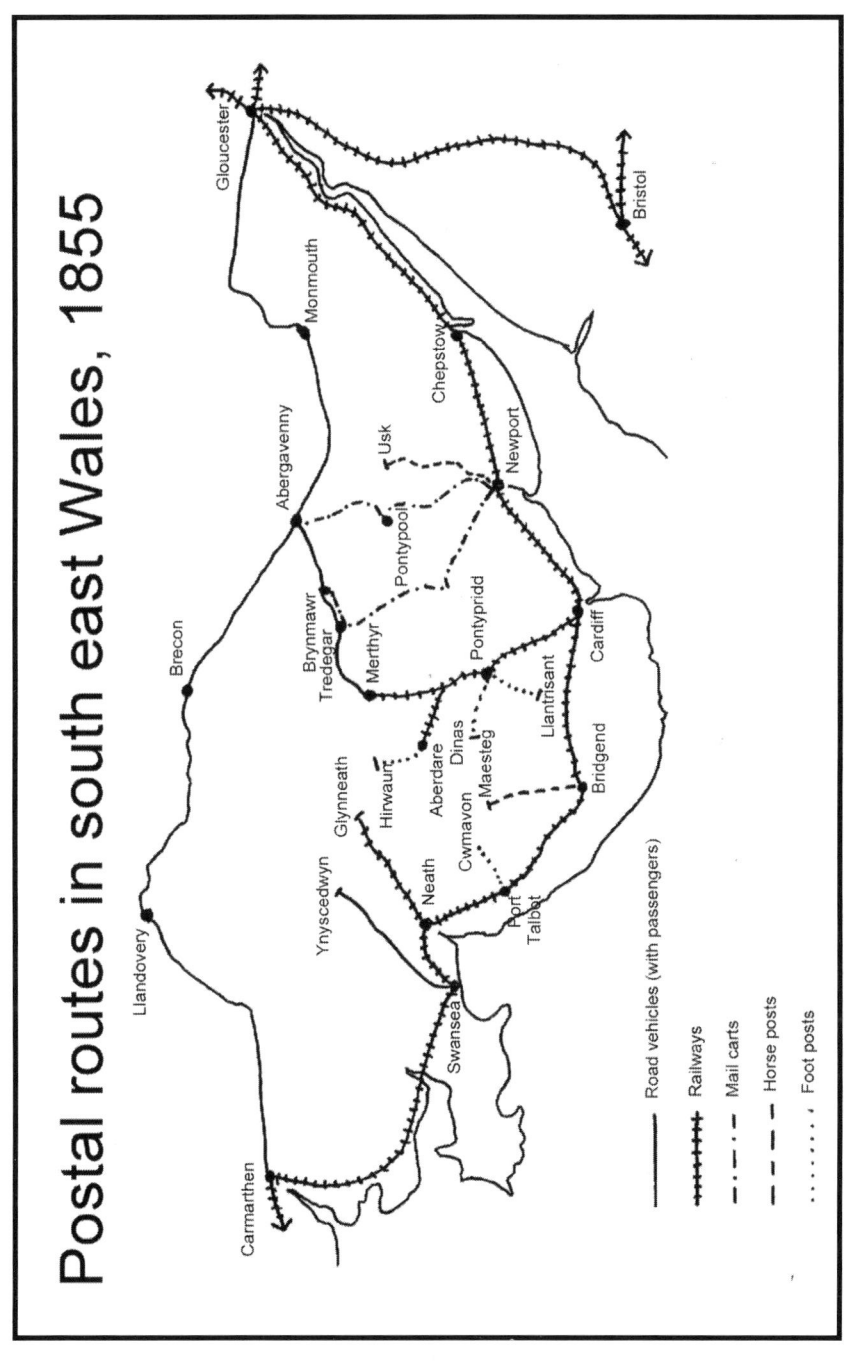

Postal routes in south east Wales, 1855

Swansea where it arrived at about 10.00 a.m. Then on 11 November the London mail was transferred back to the Gloucester route. The train left Paddington at 8.50 p.m., Gloucester at 1.35 a.m. and arrived in Swansea at 6.20 a.m. Until August 1852, when the railway bridge over the Wye at Chepstow was completed, the mail still had to be carried across the river by road but this did not cause any serious delay. With the transfer of the mails to the railway for the entire journey, south Wales finally received a postal service that can be regarded as recognisably modern: most letters arrived early in the morning in time for a delivery before the working day started, and for outgoing mail the box did not close until the end of the afternoon. Even so, more remote places which did not enjoy the benefits of a main line railway (or any railway at all) still did not receive their letters until much later in the day. Mail carts, or for a time a few residual mail coaches, continued to serve these places until the railway reached them.

Chapter 3

Letter writers

The two previous chapters have concentrated on the personnel who managed and operated the postal service. This chapter will consider its users. Who in the valleys was writing letters? What were they writing about? How many letters were they writing? Where were these letters going? What, in sum, was the nature of the demand for postal services?

The letter writers fall into various different groups and each of these will be considered in turn. First, of course, there were the ironmasters and (probably to a lesser extent) the coal owners writing frequent letters to a wide range of correspondents, including their agents, their customers and their lawyers, and to one another. Then there were the professional classes, in particular bankers and solicitors: the ironmasters gave them plenty of business and because many of them were located outside the coalfield in places such as Brecon this resulted in a steady exchange of correspondence. There were also the tradesmen and shopkeepers whose businesses supported and were in turn supported by the industrial communities in which they were located. There were the social letter writers, particularly among the landed classes and the gentry, including the families of the ironmasters. Perhaps most interesting of all there were the ordinary people, the *gwerin*, who have left little evidence for their use of the post, but who probably used it to a greater extent than is commonly believed. Finally there are the newspapers which formed a large and bulky component of every mail and whose arrival was eagerly awaited in every community throughout the land.

The ironmasters

Undoubtedly the requirements of the ironmasters was the most important consideration in the provision of postal services. The industry was structured in such a way that the iron companies were necessarily dependent on an efficient postal service and that they generated an extensive and wide-ranging correspondence, both inwards and outwards. The land-locked location of the ironworks was ideal for the production of iron but far from ideal when it came to marketing. The manufacture of finished goods (with the notable exception from the late 1820s of rails) never developed as an ancillary to the furnaces: south Wales simply produced bar iron, its speciality, which was shipped to merchants and manufacturers outside the region. In order to facilitate this trade the iron companies appointed local agents throughout the country, the most important markets being in London and Bristol. The role of the agents was to receive orders from customers and transmit them to the ironworks. In turn the iron companies preferred to deal with these agents rather than directly with the end user. The larger iron companies also had sales offices in more accessible locations. Thus Dowlais had its London House and its Cardiff Agency and the Crawshay family had both the Cyfarthfa works in Merthyr and the London Yard in London (although relations between the works and the yard frequently became strained). Whatever the arrangement, it necessitated a great deal of communication between the ironworks and the sales agents to minimise the disadvantages imposed by the location of production. The inter-dependency of production and communication is brought out by Samuel Woodcock, the Post Office surveyor, in a report of 1808: ' ... the Iron Manufactories & other Concerns in this part of Wales ... are continually increasing their Business and correspondence'.

The iron companies all had their wharves in Cardiff or Newport with a local shipping agent whose responsibility it was to arrange for the despatch of orders, and this too was a source of correspondence. Thus, when the question of an official post between Newport and Pontypool was under consideration in 1814 Woodcock could report that a considerable correspondence existed between Abergavenny and Newport. Since Abergavenny was the post town used by most of the Monmouthshire ironworks, much of this correspondence must have been letters between the ironworks and their shipping agents. This is shown clearly in 1832: John Moggridge, a coal owner in Blackwood, wrote to Freeling to urge the establishment of an official post between Tredegar and Newport where ' ... all of them [i.e. the ironmasters] have large wharf and shipping establishments with which they are in daily communication'.

Another frequent flow of correspondence was between owners who were resident in other parts of the country and the local managers of the ironworks. William Taitt's many worried letters from Cardiff to successive managers at Dowlais and Richard Crawshay's letters from London to James Cockshutt in Cyfarthfa have already been mentioned. In the following generation William Crawshay I, also in London, wrote frequently to his son, William II at Cyfarthfa, generally to find fault with the way in which he was conducting the business. In an earlier period and in the opposite direction there are the letters sent by Charles Wood during the construction of Cyfarthfa in 1766 to report progress to his principals, Anthony Bacon and William Brownrigg. There is no reason to suppose that similar flows of letters did not pass between William Thompson and William Forman in London and the managers of Penydarren, or between Forman and the managers of the Aberdare, Tredegar, Bute and Rhymney ironworks, in all of which he had an interest.

Because of the importance of the iron trade to the regional economy the ironmasters expected their preferences to be taken into account in the provision of postal services. When an official post from Cardiff to Merthyr was proposed in 1804, William Crawshay took it for granted that Freeling should be guided by his views and the views of his fellow ironmasters. His preference was for a 3-day post: 'it seems that this was not at first the general desire of the Iron Masters', reported Woodcock, 'tho Mr. Crawshay says he has converted them to his opinion'. Be that as it may, the post was nevertheless set up as a 5-day post. This was not the only occasion on which Crawshay was out of step with his fellow ironmasters. Josiah John Guest of Dowlais also frequently took the initiative in urging improvements on the Post Office. In 1821 he presented a petition from Merthyr requesting the upgrading of the horse post to a mail coach; it was signed by the three iron companies, '(Mr. Crawshay alone objecting)'. In 1838 Guest campaigned for a mail coach from Abergavenny to Birmingham and in 1846 for the transfer of the Merthyr mails from a mail cart to the Taff Vale Railway. However, in contrast to Guest's requests for the provision of better facilities by the Post Office there was the continued resistance on the part of Samuel Homfray of Tredegar to any attempt to provide an official service in the Sirhowy and Ebbw valleys in place of his own profitable private post to Newport. The fact that the Post Office, under both Freeling and

Maberly, allowed any attempt to create an official post to be thwarted by Homfray is a blatant instance of the influence that the ironmasters had in the provision of postal services. The efforts made by the Merthyr ironmasters to ensure that the official postal services met their requirements and Homfray's fierce defence of his own postal system both indicate, in different ways, the ironmasters' determination to exert control over this aspect of their businesses.

It is fortunate that a number of collections of letters deriving from the iron companies survive, either original incoming letters or copies of outgoing letters. The most notable of these are the Dowlais collection in the Glamorgan Record Office and the Cyfarthfa papers in the National Library of Wales. The letter books of Richard Crawshay of the Cyfarthfa ironworks and of the Ebbw Vale Iron Company, both in the Gwent Record Office, are also of much value, as is the letter book of Richard Hill of the Plymouth ironworks in the National Library. Letters to or from the ironmasters are also preserved in other collections, such as the Maybery, Bute Estate and Tredegar archives, again all in the National Library.

The Dowlais letters

The Dowlais letters comprise an amazing total of 563,000 documents covering the period from 1782 to 1917. They fall into four groups – copy letters despatched from Dowlais works (which have only survived for 1782-94 and 1874-83), incoming letters received at Dowlais, letters from the Cardiff Agency received at Dowlais, and letters from the London House, also received at Dowlais. A selection edited by the then Glamorgan County Archivist, Madeleine Elsas, was published in 1960.[1] She arranged her material into seven chapters, each with a number of sub-headings. The contents page gives an idea of the vast range of the collection: The Ironmasters, Masters and Men, The Business, Markets and Sales, Transport and Communications, Technical, and Politics.

The earliest of the surviving documents is the Outgoing Letter Book of 1782-94[2] and at this period the number of letters sent out from Dowlais is not high. The volume starts in July 1782 and for the last six months of that year 57 letters were copied into the book, an average of 9.5 per

---

[1] Elsas 1960
[2] Glam RO DG/A/2

month. Until 1793 this average is exceeded in only one year (1784) but after that there is a marked increase. In 1793 the monthly average was 12.08 and in 1794 it had increased again to 18.67. The reasons for this are not fully clear. It might simply indicate a more systematic ethos in the office at Dowlais which resulted in copies being made of letters which might not have been recorded at an earlier date. It is perhaps significant that the increase starts in 1792 when Robert Thompson replaced the rather unsatisfactory Thomas Guest as works manager.

The contents of the letters generally relate to orders and sales, including shipping arrangements. Letters were sent to addresses in all parts of the United Kingdom, especially to London, Bristol and Dublin. With the rather surprising exception of Swansea, addresses in south Wales were well represented although there were no frequent correspondents. Many of the letters were to the company's agents in different parts of the country through whom orders were placed. Among these were regular agents in Manchester, Newcastle, Gateshead and Hawarden, Flintshire (who also covered Liverpool). In the London market Dowlais products were sold both through agents and directly to the foundries. In Bristol they again sold either directly to the foundries or to merchant houses. In Ireland merchants in Dublin, Cork and Waterford acted as agents for sales in that country.

The Main Series Letter Books, or the collection of incoming letters, starts in 1792 and continues to 1917.[3] John Guest, who had been appointed manager in 1767 and had subsequently acquired a controlling interest in the company, died in 1787; his interest in the company was left to his sons, Thomas and John, and his son-in-law, William Taitt. Thomas Guest became manager of the ironworks while Taitt controlled overall policy from his base in Cardiff, where he was also responsible for sales. He was indisputably the dominant partner until his death in 1815. Most of the letters in this collection prior to that date are from Taitt to Thomas Guest or to one or other of his sucessors as works managers in Dowlais. Taitt's 'commanding, curt, and even rude letters to everyone, and especially to his brother-in-law, make fascinating reading and give proof of his energy, industry and technical

---

[3] Glam RO DG/A/1

knowledge', as Madeleine Elsas puts it.[4] He was a great worrier who concerned himself with every detail of the company's affairs. He wrote frequently, often several times a week, and the arrival of the postman must often have caused hearts to sink at Dowlais. Nevertheless it was thanks to Taitt that Dowlais developed from being the small and rather backward concern that it had been in the 1790s into one of the largest ironworks in south Wales.

During the period in which Taitt controlled the business from Cardiff the average number of letters arriving at Dowlais ranged from about 10 to 18 a month, by far the greatest number coming from Taitt himself. However, from 1816 onwards this number shows a marked increase., which must indicate that business that had previously been handled by Taitt at Cardiff was now being directed to Dowlais for the attention of the young John (later Sir John) Guest who had assumed control of the company on the death of Taitt. During the period 1792-1815 the highest annual number of letters received at Dowlais was 279 in 1812 (some of which were carried privately). In 1822 no less than 1,840 letters were received of which perhaps 100 to 150 were not handled by the Post Office.

Letters from Taitt to the managers in Dowlais concerned all aspects of the operations, including supplies of coal, ore and timber, the placing and fulfilling of orders, haulage, quality control, company property, the truck shop, labour, finance, legal issues and personal matters. Other letters received at Dowlais were from other ironmasters, iron merchants, customers, hauliers and carriers, solicitors, small suppliers and contractors, and employees. The letters from after 1815 give a better idea of the extent of Dowlais's correspondence, and consequently of the wide range of postal arrangements on which their business depended. To take the year 1822 as an example, letters originated from every part of the United Kingdom with England accounting for about 40 *per cent* of the total, Ireland for over 20 *per* cent, and Scotland for nearly 12 *per cent*. Interestingly the amount of business that was conducted with firms in Wales was small. If the letters from Cardiff (which generally relate to shipping arrangements) are excluded, only 7.25 *per cent* of the correspondence originated from within Wales.

---

[4] Elsas 1960, viii

There were also 53 letters from France, five from Germany and two from New York.

The next group of Dowlais letters are those sent to Dowlais from their Cardiff Agency.[5] They cover the periods 1819-25 (but with some gaps) and 1848-93. The agency was the office at the Dowlais wharf in Cardiff which was concerned with making arrangements to ship out the orders. The letters report principally on the progress of these arrangements, on the chartering of ships and on their arrivals and departures, and on the receipt of supplies for the works including oats, barley, beans, gunpowder and tallow. The wharf maintained a stock of iron on which it was able to draw but it was not always possible to meet orders in this way and letters had to be sent to Dowlais to chase up orders because customers and vessels were waiting. Bills of lading are frequently transmitted back to Dowlais for reconciliation against their records of iron supplied to the wharf. From the content of the letters it can be inferred that a corresponding stream of letters passed from the works to the wharf reporting on the progress of orders. On average during the 1820s, about 20 to 25 letters a month were despatched from the Cardiff Agency and one can guess a similar number were sent in the opposite direction.

The final group of letters are those from the Dowlais Iron Company's London House.[6] The series runs from 1839 to 1867 and with few exceptions comprises copy letters from the London House to various officials in the Dowlais works. The subject matter is largely financial, including orders for rails and other products, shipping arrangements and quality issues, the purchase of supplies for the works and the state of the iron trade and the economy in general. During the early years of this series between 200 and 300 letters were generated each year but this number starts to rise from about 1851. It is interesting that during the 1840s there is no real increase in the annual number of letters, taking one year with another. There is no discernible tendency to write more frequently as an immediate result of the reductions in postage of 1839 and 1840 which shows that commercial correspondence was hardly price sensitive: it was a necessity and the charges were just one more cost that had to be taken into account in conducting the business.

---

[5] Glam RO DG/A/4
[6] Glam RO DG/A/3

Crawshay letters

The Dowlais collection is complemented by two collections of correspondence deriving from Cyfarthfa. The first of these is the letter book of Richard Crawshay[7] which contains copies of 594 letters written by him between 1788 and 1797. There is reason to believe that only selected letters were copied into the book and therefore it does not represent his total output during this period. Crawshay started life as an iron merchant in London. In 1786 he took a lease of the Cyfarthfa ironworks following the death of its original owner, Anthony Bacon. Crawshay remained in London until the spring of 1792. He then moved to Merthyr to supervise the works in person. His son, William Crawshay, took over management of the London business. A close (but not always harmonious) connection existed between the London and Merthyr ends of the business: Cyfarthfa produced and London sold. All orders were placed in London and the iron was shipped from Cardiff, where the shipping arrangements were in the hands of Henry Charles who acted as shipping agent for all the Merthyr ironmasters.

Not all the letters that Crawshay sent during the period 1788-92, when he was still in London, were directed to Merthyr. Of those that were, by far the most were to his partner and local manager, James Cockshutt, and often concerned Cockshutt's attempts at introducing and making a success of puddling, the process patented by Henry Cort to convert cast iron into wrought iron. While Crawshay was still in London, the number of letters written to Merthyr each year ranges from 34 (1789) to 58 (1791). After he had moved to Wales in 1792 all his letters originate from Merthyr but there is a marked reduction in the number and it seems that letters were no longer copied into the book systematically. Fewer letters from this period relate to the ironworks and more relate to county affairs as Crawshay tried to establish himself in his new milieu.

Richard Crawshay remained in Merthyr until his death in 1810, by which time Cyfarthfa had become the largest ironworks in Britain. The concern then passed to his son, William Crawshay, and two other partners. Crawshay installed his son, also named William, as manager at Cyfarthfa. He himself remained in London where he continued to

---

[7] GRO D.2.162. A calendared edition was published by the South Wales Record Society (1990)

manage that end of the business and at the same time tried to control his son's activities in Merthyr. Relations between father and son were strained, not least by the fact that the interests of the London yard and Cyfarthfa ironworks did not necessarily coincide. William Crawshay I died in 1834 and his son inherited both Cyfarthfa ironworks and the London yard.

Letters and letter books of both William Crawshay I and William Crawshay II survive in a large deposit of documents in the National Library of Wales.[8] The first letter book contains copies of 288 letters from William Crawshay I in London and covers the period 1813-17, although far fewer derive from 1815-17 than from the earlier years. They are addressed to a range of correspondents, most of them in south Wales. Of the 126 letters of 1813, for instance, 44 were to William Crawshay II at Cyfarthfa and 32 were to his brother-in-law, Benjamin Hall, the owner of Rhymney ironworks and one of the partners in Cyfarthfa. Altogether 51 of the 126 letters were to recipients in or around Merthyr and a further 42 to other parts of south Wales. Similarly of the 106 letters of 1814, 43 were to Merthyr and a further 25 to south Wales.

Topics covered by the letters include operational and financial matters relating to Cyfarthfa and the state of the iron trade in general. Relations between Cyfarthfa and the London yard are a constant issue. So too are Crawshay's repeated instructions to his son to ensure that he keeps his costs down. Other topics that appear from time to time are labour relations, road and canal affairs, land purchase, rents owed and owing, supplies for the truck shop, and – just occasionally – personal matters.

The second Crawshay letter book covers the years 1830 to 1839. The letters are nearly all copies of letters written by William Crawshay II, generally from Cyfarthfa Castle. Altogether there are 439 letters, although the numbers tail off after 1834. The most prolific year was 1831 with 106 letters.

The letters relate more to the financial affairs of the Crawshay household than to the iron business. Mention of social and domestic matters appears from time to time, but generally only insofar as they have financial or legal implications. Housekeeping matters appear quite

---

[8] NLW GB 0210 CYFHFA

often, including a good deal of mail order of food and drink for family consumption and of other household supplies. Very few of the letters were written simply to maintain social or family contacts: no doubt letters of this nature were written but it was not thought necessary to make copies. Matters relating to the business are sometimes discussed, including the quality of the product, the price of iron and the supply of coal. There are frequent letters about the modernisation of Treforest tinplate works, a project which he undertook for the benefit of his son, Francis ('Frank'), including regular letters giving advice or instructions to Frank on how he was to manage the works.

The addressees are in a wide range of places, but many of them are in London or Bristol. Some of them are local including Cardiff, Swansea, Monmouth, Usk and elsewhere, but none represents a single major stream of correspondence.

These letter books are supplemented by three boxes of loose letters received by William Crawshay II between 1819 and 1855 (with a handful of letters from earlier dates). Most of them are from his father in London but there are also a number from Thomas Bold, a solicitor in Brecon, and from his brother George Crawshay and other partners in the London yard. The average number of letters is about 50 a year but how many others failed to survive cannot be told.

The topics covered are very similar to those of the 1813-17 letter book. Operations at Cyfarthfa and at Hirwaun (owned by the Crawshays since 1819) and their finances are frequently discussed, as are relations between Cyfarthfa and London. William Crawshay II's development of the tinplate works at Treforest (of which his father disapproved) also appears from 1834 onwards. The growing market for railway iron starts to appear but only occasionally: Crawshay was not as enthusiastic as Guest about this development, which is one of the reasons that Cyfarthfa lost its pre-eminence to Dowlais. Matters in Merthyr village appear from time to time, including the unrest of 1831 and the cholera of 1832 but not in any great detail. Similarly there is virtually nothing about the construction of Cyfarthfa castle in 1825. Personal matters appear from time to time but they are clearly regarded as being of secondary importance.

The collection contains a further three volumes of letters (1851-70) and five boxes of loose letters (1856-77) but these lie outside the period of this work.

A third source of information deriving from Cyfarthfa which sheds light on the letter-writing practices of the ironmasters, but at a much earlier period, is the diary of Charles Wood.[9] In 1765 Anthony Bacon and William Brownrigg took a mineral lease and started to build a furnace at Cyfarthfa. Wood, the brother-in-law of Brownrigg, was sent in from Cumberland in 1766 to take over supervision of its construction. He kept a diary for just over twelve months as the basis of his reports to his two principals. He regularly recorded the despatch and receipt of letters. In order to put his letters into the official post he sometimes used the village postman who was already making a weekly trip to Brecon, at other times he sent his own servant.

Wood's letters were partly to do with the furnace and partly to maintain contact with his family. He frequently reported progress to Bacon, sometimes several times a week, and also wrote regularly to his wife, who had remained in Cumberland, and to other friends and family members. Sometimes too he noted when he received newspapers by the post. Wood's diary is particularly valuable for the light it sheds on the operations of an early private post (see chapter 4).

Other ironmasters

Much has been made of the Dowlais and Cyfarthfa letters because these are by far the richest surviving collections of correspondence deriving from the iron companies. But there is no reason not to believe that the other iron companies generated correspondence to a similar extent. Two surviving examples are the letter books of Richard Hill of the Plymouth ironworks at Merthyr (1786-92)[10] and of the Ebbw Vale Company for the period 1824-7.[11] Both books are the solitary survivors of what must once have been extensive series. There is also a small but interesting collection of Homfray letters relating to the Penydarren ironworks in the period 1787-8.[12]

---

[9] Wood 2001, *passim*
[10] NLW MS 15334E
[11] GRO D.2472.2
[12] NLW MS 15593E

Many pages of the Hill letter book are badly stained and some are virtually illegible. Its particular interest lies in the fact that it covers the period during which Richard Hill obtained a lease of the Plymouth ironworks following the death in 1786 of its previous owner, Anthony Bacon. Hill had been an employee of Bacon at Cyfarthfa and Plymouth. On his death Bacon left the ironworks to his sons Anthony and Thomas, but since both were under age the Court of Chancery granted a lease to Hill. Much of the early correspondence concerns the negotiations that led to this settlement, but there are also letters relating to the day-to-day business of the furnace. The highest number of letters in any one month is 22 (October 1786) but frequently it is much lower than that, especially at times when Hill was away from Merthyr in London or elsewhere. He rarely notes the address of the recipient but it is clear that many of them were lawyers or others in London. There are also a good number of local correspondents in places such as Cardiff or Brecon.

The contemporary Homfray collection comprises just 11 letters from 1787-8, early years in the history of the Penydarren company which was established by the brothers Samuel and Jeremiah Homfray in 1784. Richard Forman (1733-94), the London iron merchant who had provided financial backing, asks for reports on the quantity of iron that was being produced. Richard Crawshay, still in London at this date, agrees to buy pig iron from the new furnace, but then complains at its inferior quality. A firm of solicitors in Stourbridge (the Homfrays' home territory) submit their accounts; tradesmen in Bristol, Ironbridge and Abercarn send in bills for goods supplied to the company; letters are exchanged with Henry Charles, the shipping agent in Cardiff. The collection may be small but it is an absolutely typical illustration of the range and nature of the ironmasters' correspondence.

The Ebbw Vale volume contains copy letters which all relate to the activities of the works; many of them are merely routine, such as brief notes acknowledging payment for goods supplied or covering payment for goods received. There are letters relating to supplies for the works itself (for instance, pit wood from Abergavenny or gunpowder from Edinburgh) or for the company shop. Quite frequently there are letters declining orders. This does not indicate that the company was so successful that it could afford to turn away business, rather that it was policy only to sell through designated agents. A potential customer in Glasgow was told in August 1825 that ' ... we beg to refer you to our friends Gilbert Cowan & Co, whom we consider as our salesmen at Glasgow and cannot receive an order but through their hands. We for the same reason wish all correspondence to take place through the medium of their House'. Ireland was one of its main markets and agents were located in all the chief ports, including Dublin, Cork, Waterford and Limerick. There were regular letters to all these agents acknowledging orders, providing progress reports, or detailing shipping arrangements. The geographical range was wide, covering all parts of the United Kingdom but Bristol was nearly always the leading destination, with letters to both the Bristol House and to suppliers of truck shop goods located in the city. Surprisingly there are very few letters to the Bristol banking house of Harford, Partridge & Co who owned the Ebbw Vale works: either the local managers were given a free hand and were not normally required to consult their principals or else there was another series of correspondence which has not survived. The company also had a shipping wharf at Newport which ran its own accounts and conducted a fairly small amount of correspondence: in the first quarter of 1815 its postage bill came to £1 12s 3d and in the second quarter to £1 15s 2d.[13]

In 1824 the average number of letters sent from Ebbw Vale each month was 60. This dropped to 30 in 1825 and 34 in 1827. There is no obvious explanation for this decline. It certainly cannot be related to a drop in production and hence in sales, since output at Ebbw Vale rose throughout the 1820s. A more likely explanation is that not all letters were copied into the letter book, and this is supported by information from the Post Office surveyor, Samuel Woodcock. In 1817 he wrote that the number of letters sent by private post to and from Ebbw Vale, Tredegar, Beaufort, Sirhowy and Nantyglo ironworks in a sample month was 1,696 altogether, an average of about 340 for each works, or 170 a month in each direction. Similarly in one month in 1832 the same five works generated 1,000 letters a month of which 160 were stated to have come from Ebbw Vale.

---

[13] GRO D.2472.4

Social letter writing

The letter-writing activity of the ironmasters was not confined to matters of business. They and their families also wrote and received personal and social letters. Very few letters of this character have survived, but there is enough evidence to show that they certainly existed – as one would have expected in any case. The iron towns were immigrant communities, the first-generation ironmasters just as much as the men who worked in their furnaces. Letters were a vital way of keeping in touch with friends and families in other parts of the country.

Two early examples of this kind of letter-writing are provided by Charles Wood and John Guest. Wood's letters from Merthyr to his wife and friends in Cumberland during the years 1766-7 have already been mentioned. At almost exactly the same time John Guest arrived from Shropshire to take up the post of manager of Dowlais. Charles Wilkins paints a rather florid picture of Guest, 'a lonely, somewhat a melancholic, man' sitting on a stone in front of his one small furnace, waiting anxiously for the weekly delivery of letters:

> That weekly post, brought far away over the Beacons,
> was to Guest the red-letter day of his life. He heard all
> about his Broseley home, tidings perhaps of the coming
> wife and stalwart sons; and the newspaper ... brought him
> in touch with the busy world which seemed so far away
> from the Welsh hillside ... [14]

It may be a purple passage, but it cannot be so far from the truth.

However, homesickness was not the only reason for social letter-writing. From the outset the ironmasters tried to integrate themselves into the local gentry community (perhaps never with complete success) and this led to letters on matters of common interest. After Richard Crawshay moved from London to Merthyr in 1792 the character of the letters preserved in his letter book definitely changes. They cease to be concerned primarily with the development of the works and with the iron trade; instead they relate more to county and national affairs. His correspondents included many of the leading figures among the south Wales gentry and nobility and he also developed various peripheral

---

[14] Wilkins 1903, 43

business interests. At the same time he kept up his connections in London and plenty of letters were sent to contacts in the City.

The letters of subsequent generations of the Crawshay family are also interesting in this respect. The surviving letters of Richard's son, William I, relate almost entirely to the management of Cyfarthfa ironworks and the iron trade. However, those of his grandson, William II, give some idea of the sort of letters that must have been written by any of the ironmasters on personal and domestic matters. The overarching character of these letters from the 1830s is still financial and administrative, but it is the finances and administration of the Crawshay family and household and inevitably some social and domestic topics appear. It goes without saying, of course, that the surviving Crawshay letters are a highly edited collection: the only letters that were retained by the family were those that were seen as being of potential future importance. Many personal letters must have been lost in this way.

Dowlais provides another glimpse of social letter writing by the ironmasters and their families. William Taitt often grumbled about delays to the post: in 1804 he blamed the young Josiah John Guest because he believed that someone in Guest's household was preventing 'the girl' from getting away in time to catch the post. (This was the servant who was sent down from Dowlais to the village on post days to hand over the Dowlais letters to the Cardiff postman.) He was more specific in 1805 when he voiced his suspicions that the bag was being detained by Guest's sisters 'whose correspondence they fancy to be of more consequence than our business'. Taitt made no further complaints about delays to his post, so perhaps the sisters learnt their lesson.`

The coal owners

Compared to the ironmasters, the early coal owners have a much lower visibility. No major collections of letters survive, and certainly nothing that could be compared to the Dowlais or Cyfarthfa archives. However, this should not be taken to indicate that the sale coal industry did not generate correspondence, although it was probably on a smaller scale than that of the iron industry.

The coal trade benefited greatly from the opening of canals in the 1790s which connected the coal-yielding valleys to the coast. By 1800

Swansea was shipping about 245,000 tons a year, partly local bituminous coal, partly anthracite brought down by the newly opened canal. The Monmouthshire Canal constructed at the same time led to the rapid growth of coal export through Newport and as early as 1804 over 64,000 tons of coal were being shipped there each year. By 1830 the Glamorgan valleys too had a very respectable output. In that year the tonnage of coal carried on the Glamorganshire Canal (113,749 tons) considerably exceeded the tonnage of iron (87,367 tons),[15] although by this date Newport was shipping well over 400,000 tons of coal a year.

Thus by the beginning of the nineteenth century there was a large and expanding export trade in coal. Not surprisingly, this trade was organised on much the same lines as the iron trade. Coal raised from pits and levels in the valleys was marketed on behalf of the coal owners by agents in the ports. Alternatively, the coal owners might have their own offices at the port where orders were taken and accounts maintained. Consequently the bulk of the correspondence went to offices in the ports rather than to the collieries in the valleys. The collieries and the shipping agents needed to communicate but the level of traffic was too low for the Post Office to become involved – unless, of course, the location of a colliery enabled it to take advantage of an existing mail. In other districts where the colliery was the sole or the main source of business private arrangements had to suffice. Thus, by 1814 there was a private post from Newport to Abercarn which is specifically stated to have been paid for by the 'different Collieries and Works'. Another private post in the same valley to Tredegar and Nantyglo also served one or two collieries on its way. Similarly the Rhondda valley was served by a private post until 1850.

In 1824 the Ynyscedwyn Iron Company (located towards the head of the Swansea valley) requested an official post. Rideout, who by this date had replaced Woodcock as surveyor, investigated and reported that it would not pay; all the coal owners in the valley had their 'counting houses' in Swansea where the letters were delivered: if a post were to be set up, it would only carry 'such communication as they might wish to make with the works &c at Tawe', implying that such a flow would be negligible. In one month, he reported, only 348 letters had been

---

[15] *Monmouthshire Merlin* 12 February 1831

received at Swansea for the Swansea valley. This included letters for the Ynyscedwyn Company and for a few private individuals. Around 150 were for five colliery companies. The best that Rideout could do was to advise the iron company on setting up a private post.

Use of the post by coal owners seems to have been rather greater in Monmouthshire than in Glamorgan if one can judge by one month's figures for one locality. In 1832 Rideout reported on the feasibility of replacing the private post in the Western Valleys with an official post. He took a census of letters for the one month of May, and this showed 210 letters from the colliery settlement of Blackwood and a further 20 from Woodfield, the home of the coal owner, John Moggridge. Not all of these letters would have been on colliery business, but even so there were more than from the Swansea valley. Nothing came of the proposal and the private post continued, but that was because the ironmasters insisted on retaining the private post which they operated themselves.

If this is typical of the rest of the coalfield, then it suggests that the collieries themselves generated only a limited amount of correspondence, and that mainly with the shipping agents: the real demand from the coal industry for postal services was in the ports from which the coal was shipped – and these, of course, already had well established mails.

Solicitors

As every collector of postal history material knows very well, lawyers were among the heaviest users of the postal service. When the iron industry began there were no solicitors at all in the valleys. Legal and financial business was carried out by practitioners in Brecon and other established towns in the neighbourhood. With the passage of time lawyers started to set up in the valleys, at first in Merthyr and later in some of the other towns. But even at the end of our period only a minority of valley communities had a resident solicitor.

Powell, Jones & Powell of Brecon was the most important of the local law firms; indeed, they were probably one of the largest in the whole of Wales. The firm was set up by Walter Powell and his brother John and continued under various titles well into the 19th century. An extensive collection of papers and letters originating from their office is now preserved as the Maybery collection in the National Library of Wales.[16] It contains a tremendous variety of material, both in terms of the different types of legal work that the firm undertook and in the range of their clients. They acted at one time or another for virtually every iron company in the region dealing with both company business and the ironmasters' personal affairs. The Powells also had a direct interest in the iron industry themselves since they had acquired a one-third share in the Clydach ironworks in 1803. It was not a wise move: 'that unfortunate concern the Clydach works' was a constant source of concern to Walter Powell.

It can well be understood that such an active practice with such a wide range of clients situated throughout south Wales and beyond would give rise to a great deal of correspondence. Many incoming letters do survive but they are clearly only a very small proportion of the total. From the year 1819 only 20 letters survive, of which 13 relate to the ironmasters' business. However, among the documents in the collection is the bill submitted by the postmaster of Brecon for postal charges between 9 June and 30 December 1819. This amounted to £23 10s 10d.[17] If one works on the basis that the average postage on each letter was 7d, this produces a total of about 800 incoming letters, or 1,600 per annum. It can be assumed that each of these generated at least one matching outward letter. This rate of postage (covering a range of 30-50 miles) seems reasonable, since most of the practice's business was with clients in south-east Wales.

Overall the number of letters surviving in the Maybery collection that relate to the ironmasters' business is 1,977 spread over the years 1783 to 1861. The bulk of them are from 1793 to 1841. Within this period the average number per year is 37, with the highest being 121 in 1825. If the survival rate of 1819 can be taken as typical, then this indicates many hundreds of letters a year to and from the ironmasters.

However, it is clear that by no means all of these letters would have been carried by the official post. Until 1824 letters sent by the official post between Brecon and Merthyr were routed via Carmarthen, Swansea and Cardiff, a distance of 140 miles, and took 48 hours. The

---

[16] NLW GB 0210 MAYBERY
[17] NLW Maybery 4586

establishment of a coach from Abergavenny to Merthyr in 1824 improved the situation somewhat but it was still a long way round. Consequently many letters from Brecon to the iron district were sent by private messenger. This was explicitly stated by John Jones in 1813 when he circulated a petition in favour of a direct post between the two places.[18] There are also several references in the Dowlais letters to the use of a private messenger to carry letters to Brecon, and in 1815 David Jones of Aberdare told Powell, Jones & Powell that he was 'sending the needful by our young man'. The importance of the ironmasters' business to the solicitors in Brecon must be reflected in their repeated attempts to get a direct official post. Following the unsuccessful efforts of John Jones in 1813 a further attempt was made in 1822 when Woodcock rather dismissed it by saying that 'only the Bankers and Lawyers seem to wish for it'. However, a horse post was granted the following year when Rideout, the new Surveyor, responded in a more positive way: he expected that ' ... many Letters will be sent thro' the P.O. which are now conveyed by other direct means'. The direct mail was taken off the following year when the Abergavenny to Merthyr coach was started, but this was not seen as an adequate alternative and further moves for a direct mail were made in 1839, 1841 and 1850, in both cases with the active support of solicitors and bankers.

Besides Powell & Jones there were plenty of other legal practitioners in south Wales. In 1835 there were nine attorneys in Brecon, 10 each in Cardiff and Newport, 11 in Monmouth and as many as 17 in Swansea. Even Usk had four. However, there were but few within the valleys. Merthyr had three attorneys as early as 1795 and by 1822 this number had risen to four. By 1844 there were seven. The only other towns within the valleys where the directories record an attorney were Pontypool which had five in 1835 and Pontypridd where there were four in 1852. It is thus very clear that for legal business the valleys largely depended on the older established towns outside the region, and this in turn had implications for the level of postal traffic that their business would have generated.

Of the attorneys in Merthyr the two principal ones were William Perkins and William Meyrick. The Crawshays favoured the local man,

---

[18] Jones had joined Walter Powell as a partner in about 1810 and the firm then became known as Powell, Jones & Powell

Meyrick, rather than Powell, Jones & Powell in Brecon. Meyrick is said to have made his fortune out of the Crawshays and on one occasion to have presented William Crawshay with a bill for £20,000.[19] Crawshay in turn referred to Meyrick as 'the Lord Chancellor'.[20]

William Perkins was Meyrick's chief competitor. A delightful vignette of conditions in his office in the early 1830s exists in a little book of reminiscences by Charles Herbert James (1817-90). In later life James was a solicitor who played an active part in Merthyr affairs and was its M.P. in a couple of Parliaments but he started his working life as an articled clerk in Perkins' office. It shows very clearly the great extent to which working conditions and office hours were controlled by the postal arrangements.

> At that time the mail which came in a four-horse coach from Cardiff, arrived at about half-past seven in the evening, i.e. the letter posted in London at six on one evening was in Merthyr at half-past seven the next evening. Inasmuch as the London mail went out again at seven next morning it became necessary to write the letter answering London letters by return that night. The duty of copying all letters devolved on me. Mr. Perkins was a dreadfully slow writer and long-winded, so after he had painfully written a long letter I had to copy it in a book with my own hand, for the era of copying machines had not arrived. In this way it was half-past eight, nine, or ten at night before I got home.[21]

Bankers

As with the lawyers, so it was with the bankers: until the valley towns matured to the point of being able to sustain a range of professional services, much of the ironmasters' business was given to banking houses in Brecon or Newport, especially to the Brecon Old Bank which had been established in 1778. This explains why the bankers joined the solicitors in seeking a direct mail between Brecon and Merthyr. Later on banks started to appear in Merthyr and the larger valley towns. In

---

[19] Williams 1988, 58
[20] NLW Maybery 2613-17 (1825/6)
[21] James 1892, 26-7

some cases they were independent concerns, set up by ambitious local tradesmen, in others they were branches of banks whose main offices were located elsewhere within the region.

Bankers were heavy users of the post and depended on it for the transaction of much of their business. Naturally they needed it to communicate with their customers, as letters in the Dowlais and Crawshay collections show. They were also dependent on it to maintain contact between different banks and between branches of the same bank. One of the most important functions of the post, so far as provincial bankers were concerned, was to maintain contact with the London banks. Country banks held current accounts with these banks on which they drew in order to discharge distant debts for local customers. The London banks also performed other functions for them such as cashing the bank notes of the country banks, negotiating bills of exchange, drafts, letters of credit, etc. The agent banks were also of great value in supporting provincial banks that found themselves temporarily embarrassed.[22] So far as business with the ironmasters was concerned, the banks held their current accounts, provided the large quantities of coin that were regularly required to pay the work force, and at times of expansion or financial difficulty advanced loans which could sometimes remain un-repaid for considerable periods of time.[23]

Banks were established in Merthyr from a fairly early date. According to Charles Wilkins the first bank was set up by a Mr D. Williams in about 1770[24] but this date seems improbably early and Wilkins is perhaps thinking of the bank that was set up much later by Peirce and Williams. Richard Crawshay was actively considering the formation of a bank in 1791 although it would appear without success. In 1812 the Brecon Old Bank set up a branch in Merthyr and by this time too William Crawshay had established the Cyfarthfa Ironworks Bank. Crawshay's bank was replaced in 1813 by a new bank set up by Thomas Peirce and David Williams who were described as former cashiers in the Cyfarthfa bank, although it has also been suggested –

---

[22] Hodges 1948, 86
[23] In 1832 Anthony Hill & Co of the Plymouth ironworks at Merthyr owed the Brecon Old Bank the enormous sum of £75,752. (Roberts 1958, 38)
[24] Wilkins, 1867, 529

perhaps erroneously – that Williams was the former postmaster.[25] It was not very successful. In 1816 Crawshay had to step in with financial support to restore confidence after public alarm led to a run on the bank; it suspended payments in 1824 and finally failed in 1826. Like many provincial banks at the time, it suffered from being too heavily dependent on one particular trade, in this case the iron industry.

Many of the first banks in other valley towns were set up by industrialists. Josiah John Guest of Dowlais started a bank in 1822 known as Guest, Lewis & Co. It had branches in Cardiff and Merthyr and lasted until about 1833. In Aberdare the ironmasters James and Francis Tappenden, who were bankers in their home town of Faversham, set up the Abernant Ironworks Bank. The Tappendens and their bank were declared bankrupt in 1814 and there is no evidence for any further bank in Aberdare until 1854.[26] The same applied in Monmouthshire, too. In 1818 Samuel Homfray, the Tredegar ironmaster, and J.H. Moggridge, the Blackwood coal owner, were described as 'lately' bankers in Newport.[27] Also in Newport, the banking house of Forman & Co closed in 1826 and was replaced by a new firm opened by Alderman Thompson and Samuel and Watkin Homfray.[28] (Forman and Thompson both had interests in Penydarren and other iron companies.)

Apart from the Merthyr branch of the Brecon Old Bank, commercial banks with no formal connection to the iron industry were slow to open branches in the *blaenau*. The Monmouth and Abergavenny bank had a branch in Pontypool by 1825, but it seems to have been closed by 1829.[29] Also in Pontypool, the Newport and Pontypool Bank was in existence by 1833. The West of England & South Wales District Bank opened a branch at Merthyr in about 1835, soon followed by the Monmouthshire & Glamorganshire Banking Company. By 1837 the latter company also had branches in Pontypool and Tredegar. But in

---

[25] Williams 1988, 57

[26] For a chronological listing of banking houses known to have existed in Glamorgan prior to 1900, see Roberts, c.1980

[27] It is interesting to note this collaboration between Homfray and Moggridge. Later Moggridge was to lead an attack on Homfray's private post, as described in chapter 10

[28] Burland 2006, 18

[29] *The Cambrian* 24 December 1825 (for the branch's existence); the premises were advertised for sale in *The Cambrian* 7 March 1829

1852 these three were still the only valley towns in which there were banks: at the same date Brecon alone had five.

A collection of letters which forms an interesting example of bankers' use of the post has survived among the Dowlais papers in one of the volumes of the Cardiff Agency letter books.[30] It consists of about 140 letters sent by the Cardiff branch of Guest, Lewis & Co to the Merthyr branch between January 1824 and March 1825 at a rate of about two or three a week. They are often short and relate to day-to-day transactions between the two branches, such as exchanging documents, transferring cash and reconciling accounts. By no means all of these letters were carried by post: it is obvious from the content that in some cases they were sent by a private messenger.

The banks also used the post for valuable consignments and not only for ordinary letters. 'We send you per mail a Box address'd to Mr. G. Peirce containing £600 of silver for our credit', wrote Hugh Dunne from Cardiff to the Merthyr branch of Guest, Lewis & Co in July 1824. Thirty years later, in 1854, when the mails for Merthyr reverted to the railway after a period of being carried in a ramshackle mail cart (as described in chapter 5), the *Cardiff & Merthyr Guardian* commented that '[b]ankers will no longer tremble for the safety of their parcels'.[31]

The banks and the post office seem sometimes to have been remarkably insouciant over the transmission of what even today may be regarded as large sums of money. Hugh Dunne, again, wrote to the Merthyr branch on 5 May 1824:

> We were very much surprised to hear by your letter of
> this Evening that you did not receive a Parcel containing
> £600 of our Large Notes which we left with Mr. Bird [the
> Cardiff postmaster] as usual for you yesterday Evening in
> time for the Mail. We have now enquired about it and
> altho he recollects receiving it he cannot be positive as to
> his having put it into the right bag but thinks it most
> likely it was sent to Dowlais [i.e. to the works, not to the
> bank]. If you have not yet heard of it please to make
> enquiry and inform by return of Post tomorrow.

[30] Glam RO DG/A/4/8
[31] *Cardiff & Merthyr Guardian* 7 July 1854

We send you herewith £500 and remain, Gentlemen,
your obt. servt.

Hugh Dunne

Country banks were also dependent on the post to repatriate bank notes which had been issued by another bank but which had been deposited with them. Most country banks issued their own notes and these tended to circulate within a fairly well defined area. However, in the normal course of commerce the notes could quite often travel some distance from their home area. Local clearing arrangements were therefore evolved. 'We take each other's notes', wrote John Parry Wilkins of the Brecon Old Bank in 1832, 'and exchange some weekly, some monthly, some once a fortnight'. By about 1840 country banks were also accepting each other's cheques and these obviously had to be returned to the drawer's bank for reconciliation.[32] Several of the ironmasters, including Crawshay and Guest, issued their own notes (in effect the paper equivalent of the copper tokens that circulated in the absence of Treasury coins) and the local banks were also willing to accept these. The ironmasters were prepared to support each other's issues, but again arrangements had to be in place for the return of the notes to the issuer.

Tradesmen and shopkeepers

The day-to-day needs of the population of the iron towns were generally met by a variety of independent shopkeepers and tradesmen, although in some communities the company truck shop was the main or only source of provisions and independent tradesmen of any substance did not emerge so long as the company maintained its truck shop.

Merthyr, as the largest town in the region, naturally had the greatest number and widest range of tradesmen. Their growth was particularly rapid between 1822 and 1848 and this led to the development of a distinct central business district around the market and High Street. Pontypool came second to Merthyr in terms of the numbers of its shops; Tredegar, Brynmawr and Pontypridd also developed a healthy range of retail services. But even so, the provision of retail facilities was markedly deficient. In 1851 Merthyr had far fewer shops per head of population. than towns of comparable size in England such as York

---

[32] Roberts 1958, 40

or Wolverhampton.[33] Similarly within south Wales smaller but older towns were far better provided for. In 1835 Merthyr and Dowlais together had 77 grocers whereas Brecon had 46, Abergavenny 54 and Monmouth 36. By way of contrast to the towns where an independent trading sector developed, Ebbw Vale was a prime example of the company town: the paternalist Quaker company of Harford, Partridge & Company provided houses and shops for their workforce, and it was only after their bankruptcy in 1842, when the property passed to new owners, that Ebbw Vale started to develop as an organic town with a normal range of facilities.

It would be a mistake to think of the tradesmen in the valleys as mere corner-shopkeepers. Of course this type of establishment existed, but equally there were men of substance who ran large businesses with significant turnovers. This is brought out well in Wilkins' description of David Williams, a mercer and draper who was appointed the first official postmaster of Merthyr in 1804:

> … Mr. David Williams was the great tradesman of the place. He was rich in lands, freeholds, and, moreover, still richer in the possession of several fair daughters, one of whom became Mrs. [William Milburn] Davies, and another, Mrs. Christopher James; and for a long time Mr. David Williams, Christopher James, and Davies had the trade of the village. Mr. Davies was a man of irreproachable character – a keen trader, alive to percentage and discounts, but with a morality that was unblemished. He was also, for a wonder, free from that drinking taint which, like scrofula, seems to linger in so many of our families. For his time, which, we must remember, was an age when schools were few and costly, and scholars went to Brecon or Cowbridge for their education, Davies was considered a good scholar.[34]

In Merthyr large-scale traders such as David Williams (the 'shopocracy') were initially drawn from the original village families. They prided themselves on their origins and tried to maintain the

---

[33] Carter and Wheatley 1982, 18-19
[34] Wilkins 1867, 351

distinction between themselves and the ironmasters and the labourers in the iron industry. Later they were joined by immigrants from England who were attracted by the commercial opportunities presented by a rapidly growing district. Together, they made up the small middle class of their respective towns and the postmasters were generally drawn from their number.

Few examples of the correspondence generated by the tradesmen have survived, either in public repositories or in private hands. However, there is adequate evidence to show that their businesses were dependent on the post and that they could be counted upon to support any move for its improvement. In 1821 Guest presented a petition to the Postmaster General requesting the upgrading of the horse post between Cardiff and Merthyr to a mail coach: it was signed by the bankers, and solicitors of Merthyr 'and every shopkeeper in the town'. Tellingly, there were only 58 signatures. In Monmouthshire, it was the pressure brought to bear by David Powell, a draper from Tredegar, that finally resulted in the replacement of the highly unpopular ironmasters' post with an official post. Similarly the tradesmen were among the first to complain if things went wrong. In 1840 'A Tradesman' from Swansea wrote to the press to complain about the late delivery of his letters following the increase in business as a result of the recent postal reforms. In 1856, at a public meeting in Port Talbot, an ironmonger was vociferous in his complaints about the current postal arrangements in the town.[35]

The Ebbw Vale and Dowlais letters show that where an iron company operated a truck shop the sourcing of supplies formed an important component of their correspondence. The geographical range on which they drew was wide – Ireland, west Wales, the west of England, Scotland – and clearly the postal service was essential. There is no reason to suppose that the requirements of the independent tradesmen, who had to obtain supplies on an equally large scale, were any different. Indeed, when the establishment of a mail coach from Swansea to Merthyr was being urged in 1833-4, one of the arguments put forward in its support was that it would improve communications between Merthyr and Ireland by a whole day. This was seen as an

---

[35] *Glamorgan, Monmouth & Brecon Gazette* 1 February 1840; *The Cambrian* 25 January 1856

important consideration because of the extent to which Ireland supplied Merthyr with provisions.[36] Meeting even the basic needs for food and clothing of these large communities, inconveniently distant from any existing retail centres, was no light matter and the post was about the only means by which the shopkeepers were able to maintain contact with their suppliers, to ascertain availability, to agree prices, to place orders, to confirm deliveries, to make payments and to resolve disputes.

Newspapers

In an age without broadcasting, the telephone or the internet, the only way in which news of current affairs could be disseminated widely was by means of the printed word. Consequently newspapers formed an important part of the Post Office's traffic. In 1817 even a small place like Abercarn received 66 newspapers in a month and 254 letters; in the same year the area served by Homfray's private post around Tredegar and Nantyglo received 274 newspapers a month and 882 letters. In 1824 the Swansea valley received 81 newspapers and 348 letters a month. In each of the these cases newspapers represented about 20 *per cent* of the total number of items. They would of course represent a much higher proportion of the weight and bulk of the mail. At Merthyr, in 1838, the number of newspapers posted out in one sample week was 1,047 compared to 930 letters. In the following week the corresponding figures were 308 and 945. Overall newspapers thus made up over 40 *per cent* of the outward traffic.[37] Most of the outgoing papers would have been copies of the *Glamorgan, Monmouth & Brecon Gazette and Merthyr Guardian* which was published in Merthyr weekly from 1832 until 1841 when it transferred to Cardiff.[38] The number of incoming newspapers may have been rather less. C.H. James, in his reminiscences of life in Merthyr in the late 1820s, recalls that

> The mail coach brought all the literature into the town, a
> very few newspapers once a day, and other literature
> once a month. The bookseller of the town, first my uncle

---

[36] *Glamorgan, Monmouth & Brecon Gazette and Merthyr Guardian* 7 September 1833, 18 January 1834

[37] 'First report of the Select Committee on Postage. Minutes of evidence. Appendix 4: Return ... showing the number of letters posted ... for one week ... ' PP 1837-38 (278) 453. Also 'Second report ... ' PP 1837-38 (658) 474

[38] Rees 1959, 72-6

> Job James, and afterwards his successor John Howell,
> used to have a monthly parcel which brought all the
> magazines, books, &c., that sufficed for the then wants of
> Merthyr ... The monthly supply of literature for Merthyr
> and Dowlais was contained in a parcel about two feet
> long by one thick ... [39]

James may have been exaggerating. Directories record three booksellers in Merthyr in 1822 and four in 1830. Merthyr's middle class may have been only a small proportion of the total population, but there is definite evidence of an aware, literate, well read and well informed element in the town at this date.[40]

Until 1855 newspaper publishers had to pay a duty on each sheet of paper they used for printing. In 1797 this was 3½d, increased to 4d in 1815. Abuse of the franking system often meant that newspapers were being carried free and this was corrected in 1828 when a Treasury minute confirmed that all stamped newspapers were entitled to travel free through the General Post and the Penny Post. The tax was reduced to 1d in 1836 and abolished altogether in 1855 but publishers could choose to continue to pay 1d per paper on copies that were to be sent by post. This provision was finally abolished in 1870.

The stamp duty entitled newspapers to free carriage in the mails, but difficulties arose in the distribution of papers to places off the mail routes. Papers were left for collection at farms, inns, private receiving houses and so on and were not infrequently lost or defaced. Indeed, this is a common and recurring complaint. In 1812 John Bird, the postmaster of Cardiff, had to reassure Taitt that his newspaper had never been opened or detained at Cardiff post office. Copies of the *Merthyr Guardian* were regularly lost in the post and the publishers suspected foul play: they may well have been right, the *Guardian* (rather improbably) being a high Tory paper published in a noted centre of radicalism. One of the many complaints against Homfray's private post in the Western valleys was its high charge for delivering newspapers to private residents, to say nothing of the delays: it was

---

[39] James 1892, 26-7
[40] Williams 1988, 74

claimed that newspapers sometimes took a week to get from Bristol to Nantyglo and when they arrived, they were in a dirty condition.[41]

Newspapers depended on the post for the collection of news as well as for their distribution. Unless the editor was content to publish accounts of purely local events, such as could be obtained by word of mouth, he needed the contributions of correspondents at a distance and access to other national and regional newspapers from which he could extract material. Both of these sources clearly depended on the post. Early issues of *The Cambrian* in Swansea often had to be padded with stock items in the winter when bad weather delayed the arrival of the mail coach and hence of current news.

Distribution by post was so important to the newspaper industry that papers were ready to change their publishing arrangements in response to changes to the posts. *The Cambrian* changed its press day in 1845 following the major adjustments to the mails that were made in that year. A few years later, in 1852, the *Monmouthshire Merlin* in Newport announced that it would be going to press several hours earlier than hitherto because of the recent improvements that had been made to the Tredegar and Abergavenny mails, ' ... in order to ensure the prompt transmission of the MERLIN to those important localities ...'[42] The *Merthyr Guardian* took even more drastic action: in 1841 it transferred its operations completely from Merthyr to Cardiff in order to take advantage of the better communications that were available in the latter town.

The *Gwerin*

Finally there is the interesting question of the *gwerin* and the extent to which they used the post and the purposes for which they used it. Conventional wisdom has it that before 1840 they did not use it, partly because they were illiterate, partly because it was too expensive, partly because they had no occasion to use it. All of these suppositions are suspect. It is true that very few letters from the *gwerin* have survived compared to those from industrialists or solicitors, and it is true that the reforms of 1840 led to the lower paid making a greater use of the post.

---

[41] *Bristol Mercury* 13 April 1839
[42] *Monmouthshire Merlin* 13 February 1852

Nevertheless, there is evidence that the ordinary people of the valleys were sending letters by post before 1840.

The concept of '*gwerin*' or working class needs far greater examination than can be given here. It would be quite wrong to think of a single homogenous class, all of whom had the same outlook on life, the same ambitions, the same culture. There was all the difference in the world between the skilled craftsman who could talk to an ironmaster on equal terms (and was confident in doing so because he knew the value of his skills to the ironmaster) and the newly arrived monoglot labourer from west Wales surviving as best he could in a damp cellar with no skills and no real stake in the community. The former believed in self-improvement and supported the chapels, schools and cultural activities. They enjoyed a modest prosperity; their homes were comfortably furnished (by the standards of the day) and they were literate, at least to the extent of being able to read the Bible and perhaps a weekly newspaper.

The extent to which the *gwerin* used the post is partly dependent on the extent of literacy. It has been claimed that by the end of the 18th century the majority of the population of Wales was literate, largely thanks to Griffith Jones's circulating schools and later the Sunday schools. On the other hand, an analysis of the registers in one country parish shows that in the late 18th and early 19th centuries as many as 80 *per cent* of the population had to sign the marriage register with a cross.[43] By the middle of the 19th century the evidence shows a growing number of children in school: perhaps by then at least half the population might have had some degree of literacy and some 25 *per cent* were reasonably competent in both reading and writing.[44]

One of the main purposes for which workers in the iron districts used the post was to remit money back to their homes in rural Wales. Analysis of the population shows that the iron towns drew in labour from the surrounding areas, either on a seasonal basis when there was a lull in farming activity, or to settle permanently. That these immigrant workers remitted funds to their families and used the post on a regular basis for this purpose is shown by a newspaper report of early 1840

---

[43] A figure that comes from a parish in Anglesey, but there is no reason to suppose that it is not typical of rural parishes throughout Wales at this time
[44] Jones 1997, 34-9

which comments favourably of the improvements brought about by the introduction of uniform penny post and, in the Western valleys of Monmouthshire, by the abolition of the ironmasters' private post:

> PENNY POSTAGE. The penny postage is of immense
> benefit to the men who are employed in the ironworks of
> Glamorganshire and Monmouthshire. Many of them have
> families residing in Cardiganshire, to whom they send a
> sovereign or half-sovereign per week. This used to cost
> 1s. 6d. postage under the old system, besides 4d. for
> sending the letter to the post-town in the ironmaster's bag.
> At that time there was no post-office at Tredegar,
> Nantyglo, Beaufort, Ebbw Vale or Brynmawr, although
> these places contained 30,000 inhabitants. Under the old
> system, and until a few weeks ago, a letter from Merthyr
> to Tredegar was 10d., that is, 6d. postage, and 4d. for
> coming in the ironmaster's bag: – the distance is about
> eight miles.[45]

A semi-official money order system had been in existence since the 1790s but the cost of a money order in addition to postage was expensive and for that reason most remittances continued to be made by enclosing cash in an ordinary letter. In 1838 an official money order service was introduced which proved very popular, especially given the limited banking facilities that existed in the valleys even as late as the 1850s.[46] There was a widespread call to extend the service and a waiting list developed of offices where the demand would justify the granting of Money Order Office status. The value of this service to its users is brought out in a letter of 1855 from a Minister in Beaufort which still did not have a Money Order Office at that date:

> ... a great number of labourers work here whose wives
> and families reside in other counties, and these poor men
> are obliged to loose [*sic*] a quarter or half a day's work
> almost every week in order to go to Brynmawr, Tredegar
> or Ebbw Vale to get post office orders to send to their
> families ...

---

[45] *The Cambrian* 8 February 1840
[46] Daunton 1985, 84-6

The extent of the demand is shown in the number of money orders that were granted at Taibach in 1847 where 750 orders were made out within two months.[47] Five hundred of these were for workers in Cwmavon copper works and many of them represented remittances sent back home by Irish immigrant workers – although in view of the distressed state of Ireland in 1847 these figures might not be typical.

Remitting funds back home almost certainly represented the principal use of the postal service by the working classes within the valleys, but it was not the only one. In 1832, when John Moggridge of Blackwood was trying to organise opposition to the ironmasters' private post in Sirhowy valley, one of his neighbours wrote in support of his efforts to Francis Freeling. After detailing his own difficulties, he continued: 'There are many persons living close to me who are similarly situated, but whose correspondence from their circumstances in life is very limited'. It is not entirely clear what class of person exactly the writer has in mind, possibly the local tenant farmers, but it does show occasional use of the post lower down the social scale. Again, on 19 June 1834 Lady Charlotte Guest noted in her diary: 'There came a letter this morning, threatening to "Scotch cattle" Dowlais House on or about July 2nd, unless all Irish are discharged from the Works'. This letter is a local manifestation of widespread industrial discontent and clearly originated from one of the local workmen. At this period workers were starting to organise themselves and form trades unions which were not simply local but national in scope. To do this effectively clearly implies the use of the post. Similarly from the 1830s the various local benefit societies into which workmen paid a certain amount each week started to be absorbed into national societies whose greatly increased membership gave them a stability that the smaller local societies could not attain on their own. Again, for this to be effective it required regular communication between the central offices and the local branches. The growth of nationally organised unions and benefit societies both imply regular use of the postal service by members of the working class.

After 1840 there was a distinct increase in the volume of mail. In 1843, for instance, Dowlais received about 500 letters a week; by 1847 this figure had grown to 1,100. It is inconceivable that the iron company

---

[47] WGAS D/D LE 59 (entries extracted by A. Leslie Evans from the Taibach register of transactions, 10 May to 18 July 1847)

and the tradesmen alone could have accounted for such a rapid increase. It must have been created by a growth in private correspondence and in a community such as Dowlais this can only have come from the working population. The same can be seen in other places. In 1832 Nantyglo, Ebbw Vale, Blaina and Sirhowy together received about 500 letters a month. In 1847 the receiving house at Brynmawr (covering much the same area) received 1,200 letters a week. In part this can be accounted for by the growth of Brynmawr within this period, but it surely also indicates a much greater use of the postal service by private residents.

Ebbw Vale
Sept 11th / 48
Sir

In reply to your note I have to inform you that I am not able to meet your demand. All my affect as been sold, and I am in loging and can only have enough of food for my wife & children. Therefore I believe that if you make exequition on my body that you will only throw good money after bad.

Your most obdt servt

Thos Thomas

A working class letter written from Ebbw Vale to a solicitor in Newport in 1848 and posted in Tredegar

The extent to which the working classes used the post at this period is also shown in the reminiscences published in 1921 of an anonymous writer who called herself *Hen Gymraes* ('Old Welshwoman').[48] Describing her recollections of life in Ebbw Vale in around 1850 she recalls a 'Mr James Watts, a preacher with the Baptists. He kept a shop and I believe the post office Newtown … Being the postmaster, he often had to write letters, as people were no generally educated then.'[49] It is interesting to note that James Watts *often* had to write letters: it indicates that the working class *did* use the post for whatever reasons and that they used it quite frequently. It also shows a fairly low degree of literacy, or at least of written literacy. *Hen Gymraes* goes on to say 'I need not tell you that in the Forties my

---

[48] *Hen Gymraes* 1998
[49] Newtown was a settlement adjacent to Ebbw Vale itself which was developed by the company for its furnacemen. Since an official post office was not provided in Newtown until 1858, Mr James's post office was most probably a private receiving house

correspondence was not great and it may surprise you to know that our letters cost a penny for delivery'. This reinforces the idea that the post office in Newtown was a private one, but it also indicates that the sending and receipt of letters in what was essentially a working class community was by no means unknown.

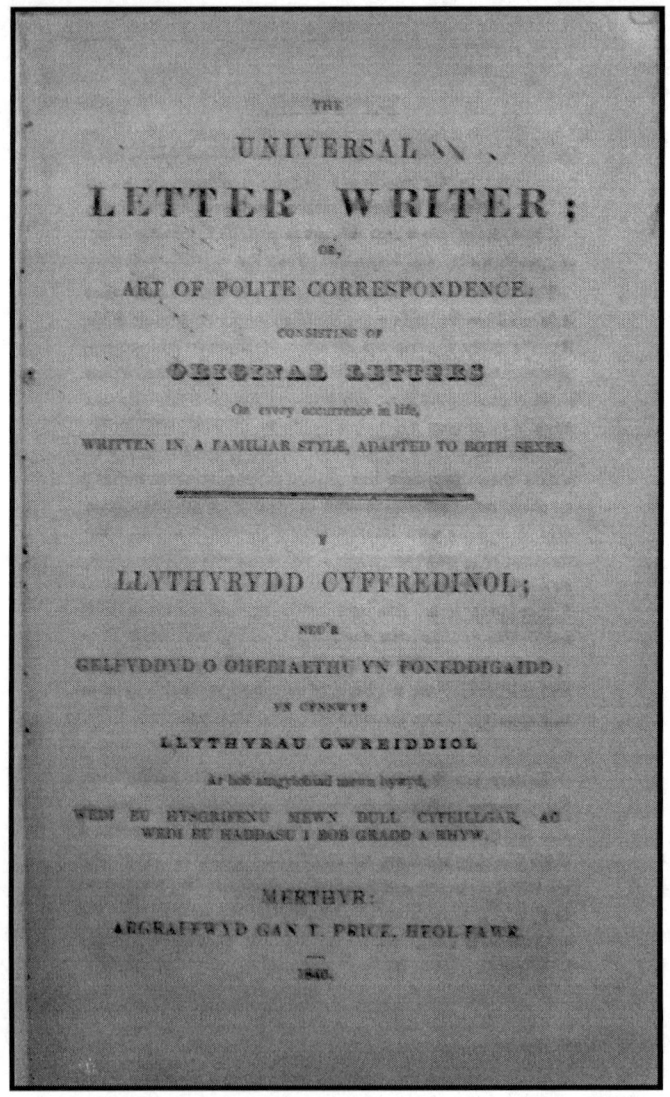

[From the collections of Cardiff University Libraries]

Some evidence for the growth of the practice of letter-writing and consequently of the increasing use of the post by the working class after 1840 can also be seen in the letter writing manuals that started to appear from the middle years of the nineteenth century in both Welsh and English. Examples are *The Universal Letter Writer* published by T. Price in Merthyr in 1840 or *Athrawydd Parod* by William Harris of Tredegar first published in 1849 and frequently reprinted.[50]

To take, by way of example, *The Universal Letter Writer*. It comprises a selection of specimen letters (parallel versions in English and Welsh) suitable for various occasions and situations. The general tone presupposes a middle-class milieu, with frequent letters 'From a gentleman' or 'To a lady', including a letter 'From a brother to a sister at a boarding school'. There is much moralising of the kind frequently to be found in other examples of improving literature of the period. The class to whom the author was principally addressing himself would seem to be the rising middle class of successful tradesmen and shopkeepers who may have come from comparatively simple backgrounds, but who wanted to know how to conduct themselves and communicate correctly in the social milieu to which they were aspiring – the class, in fact, from which the postmasters of the valley towns were generally drawn. Whilst *The Universal Letter Writer* is certainly not evidence of widespread letter-writing among the working class, there are nevertheless a few specimen letters which indicate their use of the postal service and give suggest the sort of topics on which they might have occasion to write. Among them are letters 'From a poor man to his daughter, just entered into service', 'From a parent in the country, to her daughter, a servant in London' or 'From a countrywoman with a large family, to a respectable lady in the neighbourhood, seeking a situation for one of her daughters'.

The increasing scale of emigration to the United States from about the same period, especially by skilled workers, also resulted in an flow of correspondence. Most of the examples of such letters that are known at present are from emigrants writing back to their friends and families in Wales, but presumably there was a corresponding flow in the opposite direction. The style often suggests that the writers were struggling to

---

[50] Jones 2005, 34

express themselves in an unfamiliar medium. Sometimes the letters were written by a scribe, much in the way that *Hen Gymraes* recalls the shopkeeper at Newtown doing. In many cases the author of a letter clearly intended it to be treated as a round robin: it was not written for the recipient alone, but to be passed on to other friends and family members and to be read out loud either to individuals or to groups.[51]

Whilst the available evidence is still rather fragmentary, it nevertheless seems to indicate that the working classes made a greater use of the posts than has previously been supposed. There was a marked increase from the middle of the nineteenth century following the reduction in the cost of postage and an increased level of literacy. Without question the industrialists and the professional and commercial classes were the principal users, especially in the period before 1840, but the use of the postal service by the *gwerin* should not be discounted.

---

[51] Jones 2005

Chapter 4

Merthyr Tydfil

The phenomenon of Merthyr Tydfil had a powerful hold on contemporary observers and continues to do so for present-day historians. Contemporaries were both fascinated and appalled by Merthyr, by the density of its urbanisation, by the massive scale of its industries, and by the sheer vigour that the place exhibited. Middle-class tourists visited its furnaces and forges and were vastly impressed and deliciously horrified by the noise and heat and the appearance of total pandemonium. Equally the social and economic conditions within one of the first working-class industrial mass societies, and the political tensions to which they gave rise, have resulted in a constant stream of historical writing, both academic and popular, which shows little sign of abating.[1]

Merthyr was one of the historic settlements of upland Glamorgan and until the mid eighteenth century it remained a small community in a remote corner of the county – although perhaps not quite so isolated as is sometimes portrayed. However, following the establishment of the iron industry in the *blaenau* for reasons described in the Introduction, the growth of Merthyr was remarkable and it soon became the un-disputed centre of the industry. The first furnace was established in 1759 at Dowlais, followed by Plymouth (1763), Cyfarthfa (1765) and Penydarren (1784). By 1823 24 of the 72 furnaces in south Wales were in Merthyr and 37.5 *per cent* of the total production in south Wales of 182,325 tons was produced here.

The increase in output brought about successive improvements to the transport infrastructure. The old parish road from Merthyr to Cardiff was improved by Anthony Bacon, the first ironmaster of Cyfarthfa, in 1767. It was largely replaced for iron transport by the Glamorganshire

[1] Good modern accounts of the urban development of Merthyr may be found in Williams 1988, Evans 1993, Strange 2005

Canal which was promoted mainly by Richard Crawshay, also of Cyfarthfa, and opened in 1794. As a result of a dispute with Crawshay, the ironmasters of Dowlais, Penydarren and Plymouth built a tramroad from Merthyr down to Abercynon to by-pass the upper section of the canal where frequent locks delayed navigation. This was opened in 1802. Finally in 1840-41 the Taff Vale Railway, a modern locomotive-worked railway, was completed between Cardiff and Merthyr.

Merthyr's urban growth was rapid and largely unplanned. By the time of the first census of 1801 it had become one of the largest urban areas in Wales (if not the largest). In its early days it exhibited all the characteristics of a modern third-world city – a large and fluid immigrant population attracted by high wages and living in high-density housing, unplanned, uncontrolled, and with an absence of hygiene or any basic municipal facilities. The first surge of development took place in the 1780s. By 1799 the original village around the historic parish church of St Tydfil had seen large and haphazard expansion and was well on the way to becoming fully urbanised. The uncoordinated nature of this development was particularly evident in the area of workers' housing to the south and east of the church. However, during the early years of the nineteenth century there is evidence of a more ordered development with wide straight streets in the Glebeland area to the north of the original village which was to develop into the main commercial district. At the same time there were also signs of planned industrial housing in areas such as Georgetown (now completely redeveloped) which was built to a grid-iron layout to house Cyfarthfa workers. [2]

Until well into the nineteenth century Merthyr was essentially a multi-nuclear settlement comprising the original village surrounded by the settlements that had grown up around the four ironworks. Each of the different ironworks communities fostered its own distinct identity. This was often defined in terms of hostility to the communities associated with other ironworks, an hostility which sometimes led to physical violence which the ironmasters did not necessarily discourage. The village itself contained the original families and most of the tradesmen

---

[2] Glamorgan Gwent Archaeological Trust, 'Historic landscape characterisation: Merthyr Tydfil' (accessed on 2 July 2007). Available from world wide web: http://www.ggat.org.uk/historic%20landscapes/Merthyr%20Tydfil/English/Merthyr_001.htm

and later the professionals: they maintained a distinction between themselves and the iron communities whom they regarded with some suspicion. Many took pride in belonging to old families that had lived in the village since pre-industrial days. Dowlais, too, was distinct, not only by reason of its physical location high up on the hills above the valley but also because of the paternalistic control exercised by the Guest family. The Guests were a Methodist family and this was reflected both in their own values and to some extent in the character of the community they controlled. Dowlais was perhaps rather more sober and respectful than the rest of Merthyr.

For the first fifty years of Merthyr's growth the parish vestry was the only administrative structure that existed to handle the issues which arose from a constantly shifting population living at an unprecedentedly high density. It was a system that had been devised for a small rural community and was quite unsuited to a disorganised and fluid urban society such as Merthyr had become. In a telling phrase, Merthyr has been described as 'urban but not civic'.[3] Gradually, however, Merthyr started to acquire the municipal institutions that were to give it stability and transform it eventually into a respectable county borough. In response to pressure from Richard Crawshay, who was acutely aware of the unruly state of the community, the resident ironmasters were appointed as the first local magistrates in 1792. In 1809 a Court of Requests (for the recovery of small debts) was set up. In 1822 a Select Vestry was instituted to conduct the affairs of the parish: until then participation in its running had been open to all ratepayers. In 1829 a stipendiary magistrate was appointed and in 1832 the town's first M.P. (Josiah John Guest) was elected and the first newspaper, the *Merthyr Guardian*, started to be published (to expose what its proprietors regarded as the erroneous principles of Guest). In 1834 a Poor Law Union was formed but a workhouse was not established until the surprisingly late date of 1853. The ideas of petitioning for a charter of incorporation as a borough was floated in 1837 but nothing came of it.

In the same way as Merthyr gradually acquired the administrative arrangements that were more appropriate to the size of its population, so the range of ancillary services grew and improved. The first

---

[3] Jones 1979, 15

directory to include details of the town lists about 102 'traders' of all descriptions in 1795, including specialists such as 'watchmaker', 'accomptant' and 'mathematical-instrument maker'. By 1822 the number had not increased greatly, but it is significant to note the presence of three 'booksellers, printers, &c.' Between 1822 and 1848 there was a rapid increase: the number of retail shops increased by 548 *per cent* at a time when the population increased by 165 *per cent*. Even so, Merthyr still had an exceptionally high number of people to each shop (400 in 1822, reducing to 145 in 1848).[4] The professions also started to appear. In 1795 there were already three attorneys and by 1822 there were two banks.

By 1851 a clearly identifiable central business district had become established to the north of the old village towards the top of the present High Street. Market halls were constructed in Merthyr itself in 1836 and in Dowlais in 1837. Significantly Merthyr market was a private development by two individuals, Dowlais market a company initiative, reflecting the more structured society that existed in Dowlais. Alongside the market hall permanent shops developed and housing for the tradesmen who continued to live in the district. In 1850 the shops of Merthyr were described as 'numerous, well furnished, and show all the bustle and activity of a thriving trade'.[5] This also was the district favoured by the professions and by 1851 it had taken on a rather more elevated social quality than the rest of the town.

Many of those who undertook the day to day administration through the vestry were drawn from the shopkeepers and craftsmen of the town, a very small group which was estimated in 1831 to form little more than 5 *per cent* of the total population. They were a class who benefited greatly in financial terms from providing essential services to the inhabitants of a high-wage town. As might be expected, the early postmasters of the town all belonged to this group. The ironmasters normally held themselves aloof from municipal politics and everyday administration, although their influence was maintained by the regular election of their agents to the vestry. They only became personally involved when issues arose that directly affected their interests. But eventually they felt compelled to enter fully into local affairs and from

---

[4] Carter & Wheatley 1982
[5] *Morning Chronicle* 8 April 1850, quoted by Carter & Wheatley 1982, 19

1831 onwards their influence in the Select Vestry (and above all that of William Crawshay II) was dominant.

In marked contrast to the desperate slums which characterised much of the town were the houses of the ironmasters. All were located in the immediate vicinity of their works to enable close supervision to be exercised, even if it meant noise and dirt. Dowlais House of 1817, plain and classical as befitted the rather earnest Guest family, served as the focus of the huddle of houses which grew up around the works. Penydarren House, a grand Georgian mansion built by the rather unpleasant Samuel Homfray in 1786, occupied its own comfortable grounds on a hillside opposite the works: in 1802 Malkin could say that 'the splendours of Merthyr Tydvil begin and end with this mansion'.[6] Richard Crawshay, who leased Cyfarthfa works from 1786, lived in a sizeable Georgian mansion, Cyfarthfa House, which was next to the forges, but in 1824-5 his grandson, William Crawshay II built Cyfarthfa Castle on the opposite side of the valley. Nothing illustrates the control exercised by the ironmasters more effectively than Cyfarthfa Castle. Built in a romantic Gothic style and surrounded by a park, it looked across the valley to their ironworks. Secure behind a stout boundary wall and insulated from the workmen in nearby Cefn Coed or Georgetown, the Crawshays could survey the industry from which they derived their influence and their wealth. The statement that their castle made about the location of economic and social power was totally unambiguous; its message of control was unmistakable.

As their wealth increased the ironmasters intermarried with the landed classes of the Vale and thus acquired an *entrée* to polite society. They acquired landed estates themselves and took an interest in agricultural improvement and all the other social and rural activities that went with their new position. But whilst the ironmasters may have taken part in the usual round of social activities the old Glamorgan gentry, still very conscious of the distinction between Land and Trade, never fully accepted them as equals, and in their turn the ironmasters always harboured a deep suspicion of the so-called 'Cardiff Castle set'.[7]

---

[6] Malkin 1807, 269
[7] Williams 1988, 41

Merthyr remained essentially a working-class town. However, it would be wrong to suppose that it was no more than a rough, tough frontier town. It certainly had its *Lumpenproletariat* (as the events of 1831 made very clear) but equally there was a significant element of thoughtful, cultured, well read and well informed townspeople to be found equally among the middle-class professionals, the tradespeople and the ironworkers. This culture manifested itself in the chapels, in *eisteddfodau* and above all in that strong tradition of radical political consciousness that has characterised Merthyr throughout its history.

The Brecon private post before 1786

The earliest evidence for any sort of postal service in Merthyr is a letter now in private hands dated 14 August 1742. It is written from Merthyr but the postmark is from Cardiff. The writer concludes:

> I beg you would direct to me as before, to be left at the
> Post Office in Brecon, for I shall have it sooner from
> there than from Cardiff.

There were thus two possible means of routing a letter to Merthyr. The first was through Brecon on the northern route from London to Pembroke. However, correspondents might often find it more convenient to have their letters sent to Cardiff, which was served by a by-post from Gloucester to Swansea. In either case the recipient would have had to make his own arrangements to collect the letter from the post office.

In 1759 the first of the great coke-fired furnaces was established at Dowlais followed a few years later in 1765 by Cyfarthfa of which the lessees were William Brownrigg and Anthony Bacon. Brownrigg's brother-in-law, Charles Wood, was engaged to supervise the construction of the works. He arrived at Merthyr in April 1766, and between then and May 1767 he kept a detailed diary. He often included a record of letters written and letters received and the means by which they were despatched or received.[8]

Wood always received his letters through Brecon, never through Cardiff. From the entries in his diary it can be established that the post

---

[8] Wood 2001

passed through Brecon three times a week, probably on Monday, Wednesday and Saturday, both to and from London on the same days. When Wood arrived there was already a private postman who went from Merthyr to Brecon and back on a regular basis. Almost certainly he made the return journey only once a week, on Tuesdays, out in the morning and back in the afternoon. Whenever Wood mentions sending letters by the postman it is always on a Monday – Monday, because the postman must have made a very early start on Tuesday morning. His entry for Monday 9 June 1766 illustrates this: he wrote a number of letters and then gave them to a servant ' ... to give the Postman who goes tomorrow morning for Brecon'. His entry for the previous Monday, 2 June, is similar: 'Wrote to Mr. Bacon in answer to his of the 29th ult. to be sent to Brecon by the Merthyr Post Man'. Wood must have sent a servant to Brecon to collect Bacon's letter which probably arrived there on Saturday 31 May. The fact that he was able to reply on Monday to a letter written the previous Thursday shows that the postal arrangements in these pre-mail coach days were already reasonably efficient, even at a place like Merthyr which lay well off any of the great post roads.

It was the same with incoming letters: anything that had been received at Brecon during the previous week was brought back by the postman on his Tuesday trip, but Wood had to make his own arrangements for letters arriving there on other days, unless he was content for them to remain there until the following Tuesday. Generally this meant sending a servant, as was probably the case with Bacon's letter. Similarly on Thursday 24 July 1766 he ' ... [g]ave William David 5 Letters for his son to take the Post office in Brecon tomorrow morning, & bring back what letters there may be for me etc.'. However, if on occasion an acquaintance happened to be in the town, he might pick up letters. On Sunday 1 June 1766 Wood noted that 'Captain Samuel Hughes was so kind to bring me letters from Brecon ... with three newspapers'. Hughes was the owner of an estate near Aberdare, and the mention of newspapers is a reminder that the distribution of papers was an important function of the Post before the arrival of the railway.

There is nothing in Wood's diary to indicate the basis on which the postman operated or for how long Merthyr had had a private postman. Obviously he was not in any way an employee of the General Post

Office. He may have been self-employed, or, perhaps more likely (as will be discussed below), his services were in some way contracted for by the tradesmen of the parish, the group that before the introduction of the iron industry were the only people likely to have used the postal service. He is unlikely to have been over-burdened since this group was few in number and cannot have generated much correspondence. Obviously the coming of the iron industry would have made a difference, but probably not a particularly great one at first. Wood rarely records writing more than one or two letters a week. No doubt he did not make a note of every letter, but even so he can hardly have sent more than half a dozen a week on average. One can assume that the same was true of Dowlais and Plymouth, so at this period perhaps there were just a couple of dozen letters a week going out of Merthyr and the same number coming in.

In any case the postman did not inspire confidence. 'As the postman is a drinking Man', Wood wrote on 23 June 1766, 'he sometimes loses them [i.e. letters] and it has been suspected that some of our letters have been taken from him, which must be by his consent, opened & returned ... '.

The solution to this difficulty was a private bag:

> The Postmaster in Brecon was so kind as to tell John
> Morgan our servant that if we would procure a pouch
> with a Lock & two keys & leave one with him, he would
> take care of all letters that come for us & when we sent
> any to him he would open the pouch with his key, take
> out what letters we sent and put in those that came to him
> for us, lock it & return it by the Postman.

Wood promptly asked the postmaster to obtain a suitable pouch and on 8 August he mentions for the first time sending his letters '[l]ocked up in Case'. From then on he regularly mentions despatching his letters in this way.

Very similar arrangements to those in force in 1766-7 are described by the nineteenth-century historian, Charles Wilkins. Wilkins was the postmaster of Merthyr from 1871 to 1897 and an enthusiastic, if perhaps uncritical, local historian. His account of the early postal service in *The History of Merthyr Tydfil*, obviously derived from oral tradition, is fully compatible with what can be inferred from the contemporary source of Wood's Diary:

> In early days Merthyr had no Post Office. The greatest consideration which Government authorities could be induced to give was permission for a resident of the village to intercept the messenger who carried letters from Abergavenny to Brecon ... From the messenger, the old woman employed by the Merthyr people received once or twice a week a dozen or so letters, which she brought down to the village by way of Pant Coed Ifor, delivering those for Mr. Guest to him at his furnace, and then taking the remainder to the Crown Inn which, from the fact of its being the largest public-house, and not from the name, was virtually the Post Office. Here the letters were placed on a deal table in the kitchen, and anyone who expected one was allowed to go and select his or her property, while the rest of the villagers, no doubt, exercised their intellectual faculties in puzzling out from the address the business contents of the rest.[9]

Wilkins began to collect material for his history of Merthyr towards the end of the 1850s. He claims that he gathered much information from the older inhabitants, but by that time it is doubtful whether anyone still alive had first-hand knowledge of the arrangements which he describes. Wilkins dates it to ' ... approximately ... the last quarter of the 18th century ... ' which is almost right. More precisely, it probably describes the situation that existed between the years 1767 and 1785. John Guest took over as works manager at Dowlais in 1767 and the term 'messenger' implies a period before the introduction of mail coaches in 1785. Wilkins is wrong to refer to Abergavenny: both the pre-1785 post boy and the mail coach which replaced him followed a route through Hereford and Hay to Brecon. He is doubly wrong when he says a little later that at this time ' ... it took the mail coach three days to get to Abergavenny ... ' (from London, obviously, not from Merthyr!). The mail coach, when it started to operate through Abergavenny, took little

---

[9] Wilkins 1908, 499. Wilkins 1903, 42-3 alludes to a slightly different version of this tradition, according to which ' ... a post-woman, mounted on a small pony, brought the letters from Brecon to the village of Merthyr ... '. The difference is not significant

over one day and Wilkins is simply labouring under a misunderstanding if he thought a mail coach ever took as long as three days for this journey. However, a horse post might well have taken this long, which again suggests that the arrangements which he describes relate to the period before 1785.

A very similar account appears in the writing of another local historian, presumably drawing on the same deposit of oral history. Writing in 1897, J.E. Jenkins says:

> ... all the letters addressed to Merthyr at that time were
> sent via Brecon, and carried twice weekly by a Vaynor
> female carrier, namely Mary Llewelyn (Mary the Post).[10]

There is a satisfying element of consistency between the different manifestations of the oral tradition and the written evidence of Wood. They all agree that in the period before 1786 mail was sent via Brecon where it was collected by a private contractor who carried it to Merthyr. In the 1760s this contractor was a rather unreliable postman. He seems to have been replaced by Mary Llewelyn and her pony, who can probably be identified with Wilkins' 'old woman'. Wood's evidence indicates that in the 1760s the post from Brecon only operated one day a week but by the time Mary Llewelyn took over it had been increased to twice a week.

'Pant Coed Ifor' in Wilkins' account must be the modern Pantcadifor, an outlying estate on the northern edge of Merthyr. This suggests that the route from Merthyr was through Dowlais and Pontsticill and then up the valleys now drowned by the Taf Fechan and Pentwyn reservoirs. It is then most likely to have followed a well defined track over the eastern foothills of Pen-y-Fan, in its origins perhaps a Roman road, which led due north to Brecon.

The private messenger service between Merthyr and Brecon operated outside the General Post and must have been self-funding. Wilkins states that his 'old woman' was employed by the Merthyr people, but whether that is an accurately remembered fact or just supposition on his part is an open question. Wood gives no indication at all as to how the postman of his day was remunerated. Leaving Wilkins aside, several

---

[10] Jenkins 1897, 128

funding models are possible. Mary Llewelyn may have been self-employed and made a charge for each letter she carried. However, another way of funding a local messenger service at this time was not by a fee on each letter but by a subscription raised among the leading inhabitants of the district. That this model was adopted at Merthyr is suggested by Wilkins' statement that she was 'employed by the Merthyr people' and by the absence of any mention of a charge: recipients simply took their letters. This assumes that the oral tradition on which Wilkins drew was reliable in matters of detail and there is no way of knowing whether or not this was the case. Nevertheless, it seems a quite likely scenario, with 'the Merthyr people' taken to mean the ironmasters and the shopkeepers. The fact that there was a regular postman before the advent of the iron industry indicates that the service had been instituted by the tradesmen of the town and was not a creation of the iron industry.

The Cardiff private post, 1786-1804

From 1782 another contemporary source, the Dowlais letter books, becomes available. Dowlais was still receiving its mail through Brecon in 1782, although on more than one occasion the Guests had to make this clear to correspondents who thought that their post town was Cardiff. A couple of examples are typical. On 12 August 1782 John Beeby of Dublin was told 'Youll [*sic*] please to direct for me at Dowlais Furnace near Brecon South Wales', and on 17 February 1784 William Taitt instructed Syer & King of London ' ... direct to me at Dowlais Brecon as Cardiff is not our post'.

However, in 1786 the ironmasters started to have their mail routed through Cardiff rather than Brecon. This was a decision which they chose to make themselves and not something required of them by the Post Office. It was an obvious enough choice. Since there was already a regular stream of correspondence between the ironmasters in Merthyr and their shipping agents in Cardiff, it made sense to have all their mail handled together. The introduction of a mail coach on the Cardiff road in October 1785 may also have influenced the decision.

The earliest indications of the change are in the Letter Book of Richard Hill.[11] On a letter of 25 August 1786 the address is shown as 'Cyfarthfa

[11] NLW MS 15334E

near Cardiff. However, since Hill did not normally show his address on the copies of outgoing letters it may be that Cardiff had already been the post town for some time. A little later, on 21 February 1787, he told a correspondent:

> I recd. both yours of the 6th and 15th to day by Brecon.
> We had no post last week from Brecon was the Reason of
> yours lying there. We are upon a greater Certainty of
> getting them by Cardiff if you will write on your Letter
> by Cardiff Bag.

Dowlais seems to have been a little slower in changing. They were still using Brecon in September 1786 when William Taitt reminded an attorney in London to address his letters to Brecon, 'which is well known to be our post town instead of Cardiff'. However, by March 1787 they had switched to Cardiff. The first indication of this is in a couple of letters from Taitt to George Daniell of London and his sister, both dated 19 March 1787. In a postscript in his letter to Miss Daniell he asks her 'Direct to me Dowlais, near Cardiff', and to her brother he writes 'I'll thank you for a line as soon as convenient directed by way of Cardiff'. Slightly later, writing to James Bateman of Manchester on 12 April, he explains 'We have lately changed our post you'l therefore in future direct to me at Dowlace Furnace near Cardiff'. (The spelling and the punctuation are Taitt's, or perhaps more likely, those of his clerk.) A similar message appears in several other letters of this period.

Penydarren also used Cardiff as its post town. Letters of 1787-8 in the small collection of material relating to the early years of the concern regularly show letters addressed to Jeremiah or Samuel Homfray at 'Merthyr near Cardiff' or variations on that form.[12] However, old habits obviously died hard and as late as December 1788 an iron founder of Ironbridge could still write to 'Mr Jere. Homfray & Co, Merthatidwell, Brecon' and even as late as July 1790 a Dowlais customer was still using the Brecon address.

As a result of the post town being changed a private post to and from Cardiff was set up. It was almost certainly established in 1786 and was definitely in existence by 1792, as is clear from the evidence of the incoming Dowlais letters which commence in that year. Many of the

---

[12] NLW MS 15593E

letters were written from Cardiff by William Taitt to Robert Thompson or Thomas Guest at Dowlais and he often uses expressions such as 'by the next post', 'per Monday's post' or 'let me know per Post', which all clearly demonstrate the existence of a regular private post between Cardiff and Merthyr. None of the letters has any kind of postal marking, either handstruck or manuscript. The Cardiff post is also mentioned in the *Universal British Directory* of 1795, the earliest known directory to include an entry for Merthyr, albeit a very brief one. The private post to Brecon was discontinued but a carrier's wagon still operated once a week which claimed to carry letters. In practice, it seems that the ironmasters made their own arrangements for letters to and from Brecon.

In 1787 it seems that a short-lived attempt may have been made to give Merthyr an official post. The only evidence for this is a single reported example of a handstruck mileage mark dated 6 October 1787. It is of the type issued between 1784 and 1789 and shows the distance to London as 188 miles, which indicates that the distance was measured via Cardiff, not Brecon or Abergavenny. Only one example of this mark has been recorded[13] and its present whereabouts is not known. However, there is no doubt that until 1804 Merthyr was served by a private messenger from Cardiff, and when an official post was established in 1804, it was described in the Post Office's own records as the first official post to serve the town. If the existence of this mileage mark can be confirmed, it must indicate that an attempt was made to set up an official post in 1787, but it can only have been a very short-lived attempt which has left no other trace in the historical record. It is unfortunate that this was before the first newspapers started to be produced in south Wales and before the regular series of official records begins.

According to the *Universal British Directory* of 1795, the Cardiff post operated three days a week. It does not state which days, but it is clear from the Dowlais letters that it was Monday, Wednesday and Friday. The same source also shows that the postman expected to leave Merthyr at about 2.00 p.m. although he might be delayed, sometimes until as late as 4.00 p.m., to suit the convenience of one or other of the

---

[13] Archer 1970, 81

iron companies. He arrived in Cardiff in time to meet the down mail from London which was due in Cardiff at 9.00 p.m. He handed over his outgoing letters to the postmaster in Cardiff and collected the letters that came in on the coach from London and also those that had arrived by the up mail from west Wales that morning (due in at 4.30 a.m.). He then had a short rest and started back to Merthyr where he was due at 9.00 a.m. This is consistent with a later oral tradition which has it that the ' ... mail bags were sent three times a week by "Tom", the post-boy, who rode on horse-back; he went up one day and down the next'.[14] As well as carrying letters between Merthyr and Cardiff, it is clear that the postman also collected and delivered *en route*: in November 1793 Robert Thompson wrote from Dowlais to William Vaughan of Pentyrch forge: 'It is very strange how my letters to you miscarry. I always give orders that the post Boy should leave them at Porto Bello' (evidence too for the existence of another unofficial receiving house).

The Cardiff post also appears in Samuel Woodcock's summary of c1790,[15] where 'Myrthir Tidvil' was among the places listed as having a delivery from Cardiff, 'by private messenger' understood even though this is not stated explicitly. The down mail coach from London was shown as arriving at Cardiff at 9.00 p.m. and the up mail as leaving at 4.00 a.m., times which correspond closely with those in the 1795 directory. The mail coach operated six days a week whereas the Merthyr post was only three days. Letters arriving by coach on non-post days must therefore have been held in Cardiff and delayed by 24 hours. Woodcock also lists a private messenger from Brecon to 'Mirthyr Tidvil'. This may be the Tuesday carrier listed in the *Universal British Directory*.

Not all of Taitt's letters were carried by the postman. Sometimes it is clear that he has made his own arrangements for a particular letter but this is unusual enough for him to mention it. One of his recurring themes is that of delays to the arrival of his letters from Dowlais, caused either by the messenger being detained in Merthyr or by delays in delivery at Cardiff. Letters should have been delivered to him the day after they arrived from Merthyr, i.e. early on Tuesday, Thursday

---

[14] Winstone 1883, 67; cf Wilkins 1867, 352 who gives the name 'Twm Richards' to this messenger
[15] 'The Woodcock Papers' 1992

and Saturday, but Taitt complains that sometimes he does not receive his letters until late in the evening.

Each of the ironmasters made up their own bag: Taitt complains in June 1794 that ' ... the boy left Merthyr on Monday without the Dowlais & Penydarran bags'[16] and he frequently mentions 'the bag' in other letters. ('Boy' in this context means 'postboy' and need not imply that the messenger was a minor.) The postman probably collected the bags from Penydarren and Cyfarthfa himself: in May 1793 Taitt grumbled that Jeremiah Homfray had detained the post boy 'all day'; if he had so much post he should have sent his own servant, which paints a picture of the postman hanging about the office at Penydarren while the clerks scratch away furiously trying to finish their letters. Because of its remote situation Dowlais had to make its own arrangements for getting its bag down to the village in time, and this meant sending one of their own employees. The Plymouth bag was presumably picked up by the postman *en route* to Cardiff. Anyone else with letters to send probably took them to some particular shop or inn where the postman collected them, and these all then went into a common bag.

The post was run by a man called Bird. Taitt complains that 'Bird's man' had had to wait until 4.00 p.m. for the Dowlais bag on 2 December 1799. The same thing happened in March 1802 when Bird told Taitt that his man again waited until after 4.00 p.m. Bird can perhaps be identified with John Bird of Cardiff, who was later to be appointed postmaster of that town in April 1802. However, confidence in this identification is somewhat undermined by the fact that when an official post replaced the private post in 1804 Bird showed no interest in running it. Another possible but less likely operator is a Merthyr man named Bird from whom the Dowlais company bought a wagon in 1800. He may have been a carrier and so in a position to operate a private post as part of his business.

The financial arrangement was that each of the iron companies paid a fixed annual sum to have their letters carried to and from Cardiff. What the charge was for occasional users is not known. In 1801 Dowlais and Plymouth were each paying £7 7s *p.a.* while Cyfarthfa and Penydarren paid £13 13s. This gives a clear indication of the relative lack of

---

[16] Reproduced in Elsas 1960, 106

importance of Dowlais at this time before the efforts of Taitt and after him Josiah John Guest had led to it becoming the largest of the Merthyr ironworks. The postman carried parcels as well as letters but these were not covered by the annual fee and had to be paid for separately. These parcels could contain all sorts of objects. In July 1792 Taitt asked Thompson, 'Can you procure a Leveret for me & send on friday by the Post? We are to have company on Saturday'. In 1796 Thompson tried to send some copper plates down to Cardiff, but the postman jibbed at this: they were just too heavy and so Taitt had to send someone up specially to collect them.

The official horse post 1804-1821

The thrice-weekly private post continued until 1804 when it was replaced by an official post. By this date the 'Reports' and 'Minutes' of the Postmasters General, preserved in the Royal Mail Archive are available,[17] which makes it possible to trace the development of the postal service in much greater detail. The first entry relating to Merthyr is in 1804 and relates to the establishment of an official post from Cardiff. On 19 March 1804 Francis Freeling, the Secretary to the Post Office, reported to the Postmasters General that approaches had been made the previous year to Woodcock for a regular post between Cardiff and Merthyr. Freeling did not say who had made these approaches, but it could hardly have been anyone other than the ironmasters. Woodcock had been unable to act in 1803 for some reason but now he reported favourably on the proposal. The Postmaster General agreed that a six-day post would be beneficial and unlikely to make a loss, so Woodcock set to work to make the arrangements. In June he was at Cardiff, consulting with the ironmasters. William Taitt discussed it in a letter to Thomas Guest written on 10 June 1804:

> Mr. Woodcock is now here & wishes us to fix on some
> person as Deputy Postmaster at Merthyr – you told me
> that Mr. Williams the Shopkeeper at Pontmorlais would
> like it – if so be so good as to go to him as soon as you
> receive this & tell him that the Salary paid by the Post
> Office to a Deputy Post Master is £10 per annm but that

---

[17] POST 42 (Reports), POST 35 (Minutes). They commence in 1792 and 1794 respectively

he may receive for delivery of the letters about the
Village ½d each letter – but not from those who send for
their letters – this may I conceive make it worth his while
– the Post is to go up 5 times a Week and return the same
day as it does now. If Mr. Williams will agree to
undertake the business, take him to Mr. Crawshay to
whom I spoke on the Subject yesterday merely to tell him
that he will do it & let him promise him every
accommodation in his power – let me know by return of
Post Williams's determination so that Mr. Woodcock may
get him appointed, or if he decline, inform me who else
you would recommend. Mr Crawshay speaks well of
Williams at all events. I think he should try it for one
year.

Williams was one of the leading tradesmen in Merthyr, although this is
hardly apparent in Taitt's rather patronising description. He must have
signified his consent immediately for on 12 June Taitt was able to tell
Guest that he was going to nominate him to Woodcock as the deputy
postmaster. The title reflects that his position was as a deputy to the
postmaster of Cardiff. As well as arranging a deputy at Merthyr
Woodcock had also to find a contractor to carry the letters to and from
Cardiff. His original estimate for this had been £150 *p.a.*, but Mr
Edwards of the White Lion Inn at Cardiff, a former guard on the
Milford mail, was prepared to do it for £120. On 30 June Freeling was
able to report to the Postmaster General that all the arrangements were
in place, and on 6 July Taitt wrote to Guest that ' ... the New Post
commences tomorrow'. The official post did indeed start on 7 July 1804
and the new arrangements are immediately apparent in the Dowlais
letters. Taitt's letters from Cardiff to Merthyr dated up to and including
6 July have no postal markings, but from 8 July onwards similar letters
invariably have the official Cardiff boxed mileage mark.

The post operated five days a week, the missing days being Tuesday
and Friday. It arrived at Merthyr at 9.00 a.m. and left at 3.00 p.m.
'precisely', as Taitt emphasised to Guest on 8 July. The early arrival in
the morning implies that the Merthyr post was timed to leave Cardiff
after the arrival of the up mail from Ireland to London at 4.30 a.m. It
would also have carried the down letters which had come in from

Bristol and London on the coach which had reached Cardiff at 9.00 p.m. the previous evening. Outward letters left Merthyr at 3.00 p.m. and would have arrived in Cardiff in time to connect with the 9.00 p.m. mail to Ireland and the 5.00 a.m. mail to Bristol and London the following morning. Thus the time to get a letter from London to Merthyr was about 36 hours, assuming it did not happen to arrive at Cardiff on one of Merthyr's mail-less days. In the opposite direction it was slightly longer. The arrival and departure times were such that a reply from Merthyr by return of post could easily be managed within working hours.

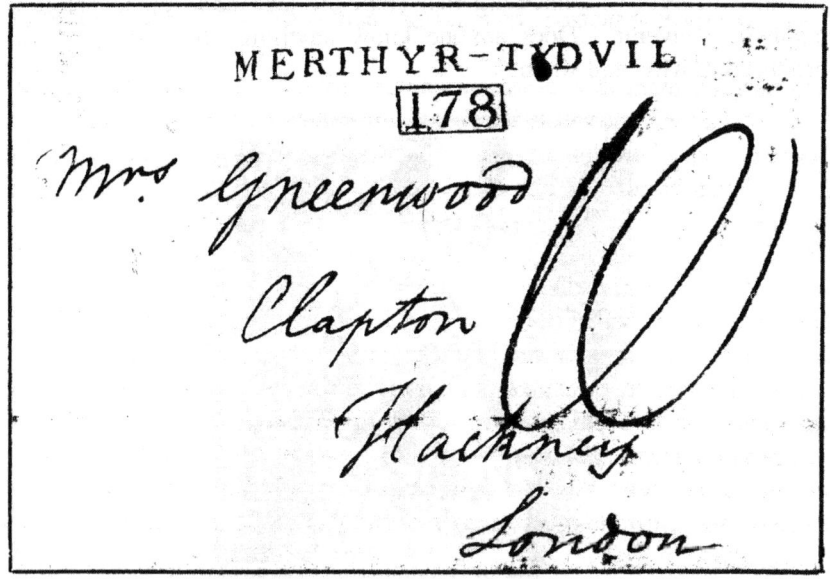

Letter from Merthyr dated 26 July 1805, just over one year after the commencement of the official post and the earliest known example of this type of mark. (By courtesy of Paul Gaywood)

In 1814 a 'Report of Riding Work' made by Woodcock showed similar but slightly improved timings: the Merthyr mail left Cardiff at 4.00 a.m. and in the opposite direction was due into Cardiff at 7.00 p.m., which indicates that the departure from Merthyr was still at 3.00 p.m. Woodcock's report for 1820 showed that the departure times were 4.00 a.m. from Cardiff and 4.00 p.m. from Merthyr, the journey taking four hours in each direction.

When the question of establishing an official post had first been raised, Woodcock had mentioned the possibility of a guarantee against any loss on the service; in other words, a Fifth Clause post was considered. This seems to have alarmed Richard Crawshay who wrote to Freeling to request that the post be made three days only. 'It seems that this was not at first the general desire of the Iron Masters', reports Freeling, 'tho' Mr. Crawshay says he has converted them to his opinion'. Whether he had or not, Crawshay had no reason to fear that he might be called upon to bail out the General Post Office, for the new post was to be on the regular establishment. 81 had 'no doubt that the produce will cover the expence' and his expectations were fully justified.

The post was made experimental for the first year, but by the end of this period, in the twelve months to July 1805, it had produced a surplus of £300 and became permanent. Partly this was due to the general increase in postal rates of March 1805, but much of it was due to new business. For instance, the so called 'short letters' between Cardiff and Merthyr were pure gain to the Revenue since before the official post was established they had been carried by private contractors and had brought in no income to the Post Office. In August 1805 the Postmaster General noted that ' ... it gives me satisfaction to find the increased activity of Wales', and in July 1806 Freeling observed that ' ... the establishment from Merthyr to Cardiff has proved to be a very productive one' – as well it might, serving the centre of the largest iron-smelting district in the country which was enjoying boom conditions because of the demand for military materials brought about by the Napoleonic wars. The surplus of £300 in 1804/05 increased to £624 in 1810/11.

The level of business seems to have come as rather a shock to the new Deputy, David Williams. By July 1805 he was talking about resigning. 'If he means to give up the business he must inform the Post Office', wrote Taitt.[18] In the same letter he noted that 'Williams is not to deliver or receive letters in church time, i.e. between 11 and 1 o'clock'. By December 1805 Williams had resigned and his place had been taken by William Milbourne Davis, his son-in-law and partner in the drapery business and an equally prominent member of Merthyr's trading class.

---

[18] Glam RO DG/A/1 (1804-05), fol. 476-7

He was presumably still in post in June 1806 and so would have enjoyed the generous increase in the deputy's annual salary from £10 to £14 awarded in that year. There is a further mention of a Davis at the post office in January 1814 and this was probably the same individual.

However, while the profits of the Merthyr horse post increased, so too did the costs. Edwards, the original contractor, had undertaken the first year for £120. He had been granted the ride not only because he was the cheapest, but also because the postmaster of Cardiff, John Bird, had shown no interest in it. Freeling believed that as a general rule local postmasters ought to be responsible for horse posts in their district. However, he accepted that in the first year Edwards ' ... has done the business remarkably well ... ' and agreed to paying him £130 for the coming year, 1805/06. In July 1806 Edwards put in a request for a further increase to £140, but this may not have been granted, for by November he had given notice of his intention to surrender the contract. It may well be that he had tendered too low in the first place, for Freeling noted in 1805 that nobody else was willing to do it for less that £150 *p.a.* The Marquis of Bute, who controlled much of what happened in Cardiff, recommended that Philip Woolcot be given the contract. Freeling regarded him as unsuitable, but it seems that Woolcot may have got the contract all the same. He was the innkeeper of the *White Lion* in Cardiff. The new contractor asked for an increase from £140 to £150 in 1809 on the grounds that the cost of provender had increased. In 1810 there is a reference to 'the conditional annual Gratuity' of £10.

In 1812 an application was made to increase the frequency of the mail from five days a week to daily. In view of this Woodcock proposed to increase the allowance to the contractor from £150 to £200 *p.a.* Freeling had no hesitation in recommending that the application be granted. He anticipated that it would produce an increase of one-eighth on the existing annual takings of £624. The daily post was duly introduced. In 1816 the allowance was reduced to £175 *p.a.*, no doubt a consequence of the depression which affected all parts of the economy in the years following the end of the war. £175 *p.a.* seems to have been a hard bargain for the contractor. By 1818 he was awarded a temporary increase of 1s a day to meet the increased cost of provender and it was agreed in October that this increase should be made permanent.

Freeling noted that nobody would take on the contract without a substantial increase on the existing rates.

The Brecon official post, 1823

The official post between Merthyr and Cardiff had proved itself a success, but nothing had been done to improve the service between Brecon and Merthyr. Once Cardiff had been made the post town for Merthyr this route obviously lost much of its importance; nevertheless there was still the need for a regular link between the two places, mainly because the ironmasters' local solicitors and bankers tended to be based in Brecon. In 1813, therefore, a request was made for a daily post between Merthyr and Brecon. The initiative came from the Brecon solicitor, John Jones, who organised a petition in his own town and then approached Taitt with a request to collect signatures in Merthyr. The Earl of Camden, a prominent landowner in the Brecon area, had agreed to forward the memorials to the Postmasters General. It was claimed in the petition that ' ... the commercial intercourse between Brecon and Merthyr Tidvil is very considerable'. It was also claimed that communication between the iron districts of south Wales and Staffordshire would be improved and, interestingly, that ' ... in Consequence of the Delay & Inconvenience attending such Communication a great Number of Letters are conveyed by individuals from the one place to the other and thereby the Revenue is very much prejudiced'. This is certainly born out by the Dowlais letters where there are regular references to private messengers being sent to carry a letter to Brecon. It is understandable, given that the existing route for mail between Merthyr and Brecon was via Cardiff, Swansea and Carmarthen, which meant that a letter had to travel 140 miles to cover a direct distance of 18 miles.[19]

Despite the involvement of Lord Camden the 1813 petition was unsuccessful, as was a further request in 1819. It was reckoned that the costs of a daily service would be at least £60 *p.a.* whilst the takings would be less than £30. Freeling reported that it would be impossible to meet the request ' ... without deserting every established principle on which we have hitherto proceeded.'

---

[19] Glam RO DG/A/1 (1813, A-L), fol. 137-8; reproduced in Elsas 1960, 84-5

The question of a post from Merthyr to Brecon was raised yet again in 1822 and 1823. As on the previous occasions the main arguments put forward were the improvement of communication between places on the Gloucester-Carmarthen road and the Bristol-Carmarthen road and the increase in revenue to be derived from letters which were being carried privately. The Postmaster General was not in favour in 1822: 'not to be encouraged', he noted. However, the following year the new Surveyor for the district, Charles Rideout, proposed a ride to connect with the mail coach at Merthyr which would cost £89 2s *p.a.* but which would make a reply possible, e.g. from Brecon to Swansea, within four days rather than six under the existing arrangements. At last the Postmaster General approved and a horse post started on 8 December 1823. It is doubtful whether it survived very long since the introduction of a mail coach from Abergavenny to Merthyr in 1824 (see below) altogether changed the situation.

Cardiff-Merthyr and other mail coaches, 1821-1841

In April 1821 the Cardiff horse post was replaced with a mail coach, an enhancement which indicates the amount of business that Merthyr was able to generate for the Post Office. In January 1821 Josiah John Guest submitted a petition to the Postmasters General making this request. It was signed by representatives of three iron companies (Dowlais, Penydarren and Plymouth), 'Mr. Crawshay alone objecting'. 'Mr. Crawshay' by this time was William Crawshay II, the son of Richard, but like his grandfather he was frequently opposed to any proposal originating from the other three iron companies – and in any case there was never any love lost between William Crawshay and John Guest. It was also signed by the bankers, solicitors 'and every shopkeeper in the town', a total of about 58 signatures which gives a clear indication of the socio-economic composition of the town which by this date had a population of 17,404. (According to contemporary directories, there were more than 58 professionals and tradesmen in the town at this time but not a great many more.) The petitioners also asked that this coach should leave Cardiff as soon as possible after the arrival of the coach from London and that the post office at Merthyr should be open for a short time after its arrival. These requests were made because in July 1820 the times of the up and down Bristol mails had been speeded up and the timings completely recast. The up coach now left Cardiff at about 5.00 p.m. and the down coach arrived at about 6.00 p.m. It appears that the times of the horse post to Merthyr had remained unaltered, so letters to London were being held in Cardiff for nearly 24 hours and letters from London for about ten.

Freeling reminded the Postmasters General that

> Merthyr Tidvil is a place of considerable importance, it
> being the very heart of the principal Iron works in the
> Kingdom. By an account which I have had kept, it
> appears that the letters arriving at Cardiff for Merthyr by
> the London mail are as 5 to 1 in comparison to those
> from Ireland

and went on to recommend that if a mail coach were not approved, then the times of the ride should be changed to meet the London mails.

The account to which Freeling refers was compiled by John Bird, the postmaster of Cardiff, and still survives. It shows that during the period from 25 December 1825 to 20 January 1821 the postage on the letters arriving at Cardiff for Merthyr from 'Ireland, Swansea &c' was £10 19s 6d where as the postage on letters from 'London, Bristol &c' was £54 10s 11d. It bears out Freeling's claim and shows that a reliable and timely service to the east was far more important to Merthyr than to the west.[20]

A two-horse mail coach was duly authorised. It cost no more than the horse post, i.e. £175 *p.a.* and was horsed by inn-keepers in Cardiff, Newbridge (i.e. Pontypridd) and Merthyr. It ran for the first time on 6 April 1821 and took 3½ hours over the journey. The departure from Merthyr was at 1.00 p.m. with a return arrival at about 10.00 p.m., thus connecting with the Bristol mails. As a result of these changes the time between Merthyr and London was reduced to about 26 hours in either direction and there was still the whole of the morning to reply by return.

The Bristol mails were again re-timed in July 1822 and the Merthyr coach was altered in consequence. It now left Merthyr at 6.00 a.m. to connect with the up mail which left Cardiff at 10.20 a.m. and returned

---

[20] POST 42/107, 148. Bird's record is preserved as POST 30/230 Report 27

at 4.00 p.m. after the arrival of the down mail at 2.55 p.m. It seems to have been upgraded to a four-horse coach at about this time.

On 6 July 1824 a new mail coach service was started from Abergavenny to Merthyr.[21] Its principal purpose was to improve the service between London and Merthyr. The link between the two places was important, and the completion of the turnpike along the heads of the valleys in 1811 made it possible to introduce a route which avoided the difficulties of the New Passage and which was thus both faster and more reliable. Until 1839 it did not serve intermediate places such as Tredegar or Brynmawr. It also had the effect of linking Merthyr more effectively to the Midlands and the north of England. Under the new arrangements mail from London was carried on the Milford Haven coach which ran via Gloucester and so avoided the New Passage altogether. The North mail from the Midlands and the north connected with this service at Gloucester, thus avoiding a detour through Bristol

Rhyd y Blew Inn near Beaufort, the half-way point between Merthyr and Abergavenny. Horses on the mail coach were changed here and a messenger from Ebbw Vale met the coach to collect the company's letters (see p. 195)

---

[21] Reynolds 2006

and, again, the New Passage. Both bags continued on the down Milford coach and reached Abergavenny in the afternoon. They were then transferred to the coach for Merthyr where they arrived at 6.30 p.m. Departure from Merthyr in the opposite direction was at 6.00 a.m. The effect of this was that London was now less than 24 hours away, but it did mean late work in the ironmasters' offices if a reply was to be sent by return. The Cardiff mail continued to run as before but now it carried only the Western mail from Bristol, south Wales, the west of England, and Ireland. The recently introduced horse post to Brecon was probably discontinued and mail sent by coach through Abergavenny.

Following the changes of 1835, when the passage of the Severn was moved from the New Passage to the Old Passage, the mail from Bristol arrived in Merthyr at 4.30 p.m. instead of at 6.30 p.m. At the same time, the Gloucester mail, ' ... which all who have travelled by it know to be the slowest in England ... '[22] was speeded up and arrived in Abergavenny three hours earlier. Consequently Merthyr got its London mail at 3.30 p.m. instead of at 6.30 p.m. The Cardiff to Merthyr coach saw gradual improvements as well and by 1838 was doing the 25-mile journey in three hours at an average speed of nearly 8½ m.p.h., which for a two-horse coach over hilly terrain was a creditable performance.

Merthyr now had mail coach services to the south and the east. In 1827 a campaign was started to set up a third service to the west, to Neath and Swansea. Freeling was not impressed with the application: he suspected that it had been promoted by the coach proprietors in their own interest, not because there was any real need for it. 'We are already incurring an expence of £205 per ann for the Accommodation of Merthyr, the Correspondence of which does not exceed £1000 a year' – which was not a bad rate of return! The previous indication of the level of business derived from Merthyr was in 1812 when it was put at £624 a year. This shows an average rate of increase of somewhat over 4 *per cent p.a.* After continuing pressure a mail cart to Neath was eventually agreed to in 1834; it was upgraded to a mail coach and extended to Swansea in 1835. As well as the mail from south-west Wales, this service also produced improvements to the mails to and from the south of Ireland. A full account is given in chapter 8.

---

[22] *Glamorgan, Monmouth & Brecon Gazette and Merthyr Guardian* 28 March 1835

Early in 1838 the ironmasters and the leading tradesmen of Merthyr and the surrounding areas presented a petition to the Postmaster General through Josiah John Guest, since 1832 the town's first M.P., requesting a direct mail between Abergavenny and Birmingham. This would obviously have been in connection with the Merthyr-Abergavenny mail. It was turned down by Maberly, by now Secretary to the Post Office in place of Freeling, in a letter to Guest:

> Sir, – The Petition of the Ironmasters and others
> connected with the trades of Merthyr Tydvil, Dowlais,
> Nantyglo, Blaenarvon [*sic*], and other ironworks in the
> counties of Glamorgan and Monmouth, praying for the
> establishment of a Mail Coach between Birmingham and
> Abergavenny having been submitted to the Postmaster-
> General, I am commanded to acquaint you, for the
> information of the parties, that his Lordship does not
> deem it expedient to entertain any question of this nature
> at present, as the measure, if adopted, could only be of
> temporary duration, inasmuch as the post
> communications of the country generally must be
> materially changed in a very short time, by the various
> railroads which are rapidly advancing towards
> completion.[23]

Maberly's invocation of imminent wholesale rearrangements as a result of railway development was really no more than a fig leaf. It was to be another three years before Merthyr's North mail was transferred to rail in 1841 and that only for the Birmingham-Gloucester section.

However, a direct Brecon mail was reinstated in 1839. On 24 April an experimental mail cart was authorised and it started during May.[24] It does not appear to have been a success: in 1841 a memorial was presented to the Postmasters General from Brecon requesting a direct mail to Merthyr, which seems to imply that the 1839 service had already been discontinued. The income which it was estimated would arise from the service was put at £109 *p.a.*, about the same as the cost of providing it. Maberly did not recommend re-instatement. Yet

---

[23] *Glamorgan, Monmouth & Brecon Gazette and Merthyr Guardian* 31 March 1838
[24] *ibid* 27 April 1839

another attempt was made in 1850 when a memorial 'most numerously signed by the magistrates, bankers, solicitors, and principal inhabitants of the town and county of Brecon' was presented,[25] but with no more success than on previous occasions. Finally, in 1857 a mail was set up between Brecon and Merthyr using an existing stage coach. It lasted until 1863.

The effect of the railways

The completion in 1841 of the Great Western Railway between London and Bristol, and the transfer of the mail from road to rail, had little impact on Merthyr. Because of delays at Bristol, the Western mail still did not arrive at Merthyr until 4.30 p.m. and the London and North mails were still carried on the Abergavenny mail coach.

Another railway which opened in 1840-41 was the Taff Vale Railway from Cardiff to Merthyr. The Post Office was eager to use it but was unable to agree terms with the railway company. Instead a mail cart was put on in July 1841 to replace the coach which had ceased to run on the opening of the railway. In 1846 agreement was reached with the T.V.R. and mail between Cardiff and Merthyr was transferred to the train. This lasted until 1850 when a mail cart was re-introduced once again as a result of irreconcilable differences between the Post Office and the railway over train times. A full account of the relations between the Taff Vale Railway and the Post Office is contained in chapter 5.

In 1845 the G.W.R. opened to Gloucester and this did produce benefits for south Wales. New postal arrangements were brought in on 1 September. The London mail arrived at Gloucester at 1.45 a.m. by train. The mail for Cardiff and Swansea was forwarded immediately by road, but so far as Merthyr was concerned, the London mail had to wait at Gloucester for an hour or two until the North mail arrived on the train from Birmingham. The two mails were then carried on by coach from Gloucester to Abergavenny, and from Abergavenny to Merthyr, but arrival was now at 10.30 a.m. instead of at 1.00 p.m., and despatch was at 2.00 p.m. instead of 11.00 a.m. It was now possible to reply to an urgent letter on the day it was received. The use of the railway as far as Gloucester also resulted in an earlier arrival at Merthyr of the mail cart from Cardiff. The coach from Gloucester now arrived at Cardiff at

---

[25] *The Cambrian* 17 May 1850

7.39 a.m.; the mail cart to Merthyr left at 8.20 a.m. and probably reached its destination at about 11.00 a.m. instead of at 4.30 p.m. In the opposite direction the mail cart left Merthyr at 1.15 p.m. and was due back in Cardiff at 4.15 p.m. Going north, most of the mail carried by this cart would have been purely local, originating from Cardiff. There would have been some Newport mail brought in by the coach from Gloucester, and there would also have been the mail from Bristol and the south-west of England which had been waiting in Cardiff since the previous afternoon.

The South Wales Railway from Chepstow to Swansea opened in June 1850. From 27 July the coach from Gloucester ran only as far as Chepstow and mail was then transferred to a train. Merthyr's London mail was now taken away from the Abergavenny route and came in via Cardiff along with the Western mail. Arrival of these mails at Cardiff was at 5.45 a.m., but there was then a delay of three hours before the Taff Vale train left, and arrival of the mail at Merthyr was at 10.00 a.m., a less than startling gain of half an hour for the London mail and one hour for the Western mail. This arrangement only lasted for a few months until October 1850, when the mail between Cardiff and Merthyr was taken off the railway and put onto a mail cart because of the differences between the Taff Vale Railway and the Post Office. Even with a mail cart, arrival at Merthyr was still at 10.00 a.m.

The North mail continued to travel on the mail coach from Abergavenny. It was put off the train at Gloucester and transferred to the coach which was still running on the northern road as far as Abergavenny, Brecon and Llandovery. From November 1852 the Abergavenny-Merthyr coach operated on an exemption-only basis: the contractors received no payment for carrying the mail, but were granted exemption from tolls. The time-table was completely re-cast, with the coach now arriving in Merthyr at 4.30 p.m. and departing at 7.15 a.m. By 1858 the North mail had been transferred to rail and routed via Cardiff: it now arrived at 11.15 a.m. The coach continued to run, carrying only local mail between Abergavenny and Merthyr and the North mail for Brynmawr, Beaufort and Tredegar, but by 1865 it seems to have been down-graded to a mail cart.[26] It ceased altogether, with the minimum of publicity, in 1866.[27]

Following the opening of the railway between Gloucester and Chepstow mail was transferred to this route from 11 November 1851. Merthyr now received its London and Western mails at 7.45 a.m. (still by mail cart) with an outward despatch at 6.30 p.m.[28] The final railway link between south Wales and England was completed in August 1852 when the bridge over the Wye at Chepstow was opened and an unbroken line of metals stretched from London to Carmarthen. This resulted in a slight improvement at Merthyr, with the London and Western mails arriving half an hour earlier at 7.15 a.m. Finally, in July 1854 the Post Office reached agreement with the Taff Vale Railway and the London and Western mails were once again carried by rail from Cardiff. Arrival at Merthyr was now 6.15 a.m. with delivery starting at 7.15 a.m. The outward mails were despatched at 5.40 p.m.

Local deliveries

In 1804 some sort of town delivery was in existence, but it was an unofficial service, offered by the deputy and probably carried out by an employee in his service. A delivery fee of ½d per letter was charged. The first appointment of an official letter-carrier within the town and the establishment of a free local delivery was in August 1837. According to Charles Wilkins (who presumably has this appointment in mind):

> The first regular postman at Merthyr was John Garnon,
> who had the distinction of being personally very like the
> Duke of Wellington. As the work increased, he had the
> assistance of a local barber, Luff Stocker.

This individual was responsible for delivery within the town, but in 1843 his load was reduced when a messenger was appointed to carry

---

[26] *The Cambrian* 3 February 1865 refers to 'the mail car' from Merthyr to Tredegar being forced to return to Merthyr by a heavy snow fall

[27] Wilkins 1867, 352

[28] These are the times which appear in *The Times*, 22 November 1851, but 6.30 p.m. for the outgoing London mail seems too late. The up train left Cardiff at 7.50 p.m. and the mail cart must have needed at least three hours to get down from Merthyr, to say nothing of an allowance for unexpected delays and time to transfer the bags onto the train

the mails to Dowlais and deliver them both there and within Penydarren *en route*. By 1848 there were two deliveries in the town, the main one at 11.30 a.m. after the arrival of the Abergavenny and Western mails and for local mail from Aberdare and Dowlais, followed by a second delivery at 2.00 p.m. for the Swansea mail, including Hirwaun and Glynneath. This was all carried out by the one letter-carrier, but by 1850 ' ... old age [had] rendered this praiseworthy individual quite unfit for his laborious duties ... '. Not only had old age overtaken him, but the number of letters had greatly increased and Merthyr was continuing to spread outwards. Residents in outlying parts of the town found that the London letters, which they should have received in the morning, were being delayed and were being delivered with the Swansea letters in the afternoon – and that even with the help of his barber! A petition to the Postmaster General was organised, requesting that a second letter-carrier be appointed. This was agreed and in March 1850 arrangements were made to take on a second postman.[29] Garnon was earning 14s a week, but the pay offered to the new carrier was only 6s, which may explain why the man appointed to this post, James Cole, resigned within a year in February 1851. His replacement was started at 7s per week, but this was increased to 12s in 1852. Garnon struggled on until 1855 when he finally retired at the age of 81. In 1854 a petition had been presented to the Postmaster General on his behalf requesting an official pension. When this was turned down a public subscription was arranged which provided him with a pension until his death in 1857.[30]

From the 1830s onwards a number of sub-offices were opened in the district and these became the responsibility of the postmaster of Merthyr. They included Aberdare (1834), Dowlais and Hirwaun (1843), Aberaman (1847), Troedyrhiw (1852), Cefn Coed (1853) and Penydarren (1857). Responsibility for Glynneath was taken over from Neath in 1836. Accounts of Aberdare, Dowlais, Hirwaun and Glynneath will be found elsewhere.

In conjunction with the opening of the sub-office at Troedyrhiw a daily official post was set up to serve Plymouth, Pentrebach, Abercanaid, Duffryn, Troedyrhiw and Ynysygored (near the later village of

---

[29] *The Cambrian* 8 March 1850
[30] Richards 2008, including extracts from the press and a warm appreciation of Garnon written at the time of his death

Aberfan). The receiver at Troedyrhiw was allowed £3.00 *p.a.* and the letter-carrier 11s a week. Another daily post served Cefn Coed and Heolgerrig. The receiver at Cefn was also allowed £3.00 *p.a.* and a further £12.00 for delivery and conveyance of the bag to Merthyr.

Postmasters

The early postmasters were all drawn from the small class of respectable and well-to-do tradesmen. Little is known in detail of their lives but what evidence there is indicates that they were prominent figures in the administrative and commercial life of the town. Wilkins' description of the first deputy, David Williams, which makes it clear that he was running a sizeable business, has already been quoted in the previous chapter. He resigned in 1805 and was replaced by his son-in-law, William Milbourne Davis (or Davies).[31] Davis originated from Breconshire and is said to have come to Merthyr in 1790. He is also said to have owned a colliery at Abercanaid at some point, but by 1795 he had entered the drapery trade in partnership with his father-in-law.[32] He took a prominent part in the affairs of the vestry and was regarded as one of the leading tradesmen of the town. Wilkins' assessment of him as an upright and educated man has also been quoted earlier. Gwyn A. Williams uses the delightful phrase *'coq de village'* to describe him. It is not known how long Davis remained in office, but by 1822 he had been replaced by a D. Davies who is named as postmaster in a directory of that year. William Milbourne Davis 'amassed a very respectable fortune in trade, and retired to Swansea just before the sober autumn of life stole over him'. He died there in about 1840.[33]

The first postmaster to make any sort of impression on the official records of the Post Office was Rhys Davies, a draper. Perhaps he is to be identified with the Rhys Davies who was described as 'Clerk at P.O.' when his baby daughter Mary was baptised on 19 November 1820. He may well have been the son of William Milbourne Davis, or at least a very close relative, since that was the name he gave to his own son who was born in 1837 (and who died as an infant in 1840). He was in office by 1828 when the duties attached to the post were comparatively light.

---

[31] For David Williams and William Milbourne Davis, see Wilkins 1867, 351 and Williams 1988, 42, 57
[32] *Universal Directory of British Trade and Commerce* 1795
[33] Wilkins 1867, 352

At that date the post office was located in Cross Street,[34] a comparatively new street to the west of High Street in an area that was developing into the central business district of the town. The street no longer exists. By 1841 Davies and the post office were in High Street itself. The reforms of 1840 led to a great increase in business: the extended responsibilities and the changes in working practices which this brought about may have been something that he did not welcome and which he was not able to handle very well. In 1841 the Surveyor was instructed to inform him ' ... that you cannot suffer the present practice of sorting letters on the open counter to continue, and he must therefore provide a proper and suitable office for the performance of the Post Office duties forthwith ... ' on pain of removal. Davies was also a draper and this instruction conjures up images of the letters mixed up in any fashion with bales of fabric and ladies' bonnets. In 1845 he was in trouble again. In May it was decided that the messenger who carried the letters from Merthyr to Aberdare should also carry out a house-to-house delivery, but by September this had still not been implemented. Davies explained that this was because the messenger was illiterate, and so unable to read the addresses on the letters he was supposed to deliver. However, he had failed to report this to the Surveyor, and had in effect taken it on himself to decide when and how the Aberdare local delivery should be made. It also transpired that Davies had been dismissing and appointing messengers as he thought fit with no reference to the proper channels as represented by the Surveyor. Davies may have been unwilling to come to terms with a reformed Post Office: perhaps he preferred the concept of the post office as an adjunct to his main business, and one which he could choose to run in the same way as he ran his draper's shop. He was warned that any more instances of unauthorised appointments would result in his dismissal.

In 1846 Davies was in arrears in submitting payments due to the Post Office and on 2 March the Surveyor was instructed to take charge of the office. This obviously produced the desired results because by 9 March Davies had paid his arrears and was reinstated. It was accepted that it was simply a case of inefficiency, not of malpractice. There was further trouble in August 1846 when he failed to report the death of the

---

[34] *The Cambrian* 17 May 1828 contains an advertisement offering for sale three houses 'situate in Cross-street, nearly opposite the Post-Office'

Receiver at Dowlais (an office under his control) for over a month. To make matters worse, ' ... in his expln. of the matter [he] has not thought proper to utter a single expression of regret for his omission'. He was to be warned that he would be dismissed for any further case of ' ... such gross neglect of his instructions'.

To be fair to Davies, he was not well and so was unable to carry out his duties as he should. In December 1846 he was given an ultimatum: in view of his ill health he had either to provide a second assistant at his own expense, or he had to resign. This seems to indicate that under normal conditions the work of the Post Office was being carried out by a staff of three, Davies himself, an assistant in the post office and the letter-carrier. Davies died on 25 May 1847 at the age of 49. In his later years he may have become a problem for the local Surveyor and the central authorities, but he was well liked in Merthyr. The notice of his death in the local paper, which has the ring of sincerity about it describes him as ' ... one of the most popular and highly respected men in the town of Merthyr. He made friends in all the walks of life and retained their friendship to the last moment of his existence.'[35] The office passed to his widow, Gwenllian Davies, following a petition in her favour signed by 'all the most influential tradesmen of the place'.[36]

Gwenllian Davies remained in post until 1851 and died in 1858. She was succeeded by a Mr Forrest. Almost at once he seems to have become involved in some sort of financial irregularity. In November 1851 he was relieved of his duties and Maberly noted:

> I am not aware of any circumstances connected with this
> case to induce your Lordships to depart from the
> regulations. I presume, therefore, you will declare
> Merthyr Tydvil office vacant, the debt not having been
> paid within 10 days since the office was placed in charge.

No more is heard of Forrest and his place was taken by William Wilkins, a bookseller and newsagent and the father of Charles Wilkins, the author of *The History of Merthyr Tydfil* which has been mentioned several times in this chapter. John Guest took an interest in the appointment, even though he was seriously ill and was to die within the

---

[35] *Cardiff & Merthyr Guardian* 29 May 1847
[36] *ibid* 26 June 1847

year: on 21 November his wife, Lady Charlotte Guest, recorded in her diary that 'he gave me orders about writing a letter for him to sign, which was to recommend a plan for the appointment of Postmaster at Merthyr'.[37] William Wilkins was appointed postmaster and Charles Wilkins, then aged 20, became his father's chief – and only – clerk. The establishment at this time was one clerk and two letter-carriers. William Wilkins died in office in 1871 and Charles became postmaster in his place.[38]

Postmasters of Merthyr

	Appointed	In office	Resigned
David Williams	1804		1805
William Milbourne Davi(e)s	1805	1814	
D. Davies		1822	
Rhys Davies		1828	1847 (died)
Gwenllian Davies	1847		1851
Forrest	1851		1851
William Wilkins	1851		1871
Charles Wilkins	1871		1898

---

[37] Schreiber 1951, 21 November 1851
[38] The career of Charles Wilkins falls outside the period under consideration, but for a brief account, see Reynolds 1994, Wilkins 2001

Chapter 5

The Taff Vale Railway and the Post Office, 1841-1854

At the beginning of 1840 mail was still being carried from Cardiff to Merthyr Tydfil by mail coach. However, the Taff Vale Railway was under construction between these two towns and its completion would obviously require a re-assessment of the situation on the part of the Post Office. Once the railway was open the mail coach would no longer be an economic proposition for its proprietors since all the passenger business could be expected to transfer to the railway. As well as being more comfortable, the train would offer a more frequent service and would do the journey in less than half the time. The mail coach ran once a day in each direction and took three hours: when the railway opened there were three trains a day each way, soon to be increased to four, and they took less than an hour and a half. The Post Office did not wish to meet the entire cost of the coach and in any case the railway offered the possibility of a faster and more frequent means of carrying the mail to and from Cardiff. The preferred option for the Post Office would be to start sending the mail by the Taff Vale at the earliest opportunity.

The first stage of the railway, from Cardiff to Navigation House (the modern Abercynon), opened on 8 October 1840 and it was opened throughout on 21 April 1841. The Post Office entered into discussions with the company with a view to transferring the mails to rail as soon as possible. But obviously they were taken aback at the charge which the T.V.R. proposed to make, since on 26 April it was minuted that their terms were not to be accepted unless they were considerably reduced. The actual level of the fee was not recorded but neither the railway company nor the Post Office was prepared to negotiate. The Post Office had to seek an alternative solution and on 31 May 1841 it awarded a contract to Bradley, the Cardiff coach proprietor, to carry the mail for £200 *p.a.* The mail coach ran for the last time on 5 July 1841 and Bradley's mail cart presumably took over the following day. If one can trust the evidence of directories, this did not result in any reduction in the level of service, for the arrival and departure times of the Cardiff mail are virtually identical in Robson's 1840 *Directory* and Pigot's for 1844.

The mail cart was still in use in July 1846 when Sir John Guest urged the Post Office to start using the railway. No doubt he was voicing the opinion of many of his constituents. Whether or not it was due to his influence, the Post Office agreed to the terms offered by the T.V.R. and discontinued the mail cart. No great financial loss was involved. The lowest tender that had been received to maintain the mail cart was £250 *p.a.* whereas the total cost on the railway was £263 19s 5d *p.a.* This was made up of £225 for the actual carriage of the mails and a further £38 or so for incidental costs, such as carrying the mail bags between the railway station and the post office at either end of the journey and for revised arrangements for Llandaff which had hitherto been served by the mail cart. The mails were probably transferred to the railway in July or August 1846.

Right from the outset the relationship between the Post Office and the railway was one of suspicion. In October 1846 the T.V.R. announced that it wanted to change the times of its trains on Sunday. The Post Office accepted the change to the 9.00 a.m. up train from Cardiff to Merthyr, but could not agree to the proposed change to the return working which, it was claimed, would cause inconvenience. This minor spat was a straw in the wind.

In 1846 the Aberdare Railway was opened from Navigation House to Aberdare. Nominally a separate company, it was in reality a subsidiary of the Taff Vale which took it over completely in 1847. In February 1848 discussions took place with a view to transferring the Aberdare mail to the railway. Once again the railway company drove a hard bargain. Their charge for providing the service was £35 *p.a.* and they were not prepared to reduce it. The Post Office reluctantly accepted their terms, since sending the mail by train meant that the ironmasters and coal owners of Aberdare would be able to reply to incoming mail on the same day. The Post Office would in any case save £36 10s *p.a.* by not having to pay a foot messenger to carry the mails from Merthyr to Aberdare, which had been the practice since 1834. On the other hand they would have to pay the Receiver at Aberdare 8d a day to collect the bags from the station, a total of £12 13s 4d *p.a.*, so overall the increased annual cost of serving Aberdare by train was £11 3s 4d a year.

Thus from February or March 1848 mail was being sent to both Merthyr and Aberdare by train. This was the quickest and most convenient method available and undoubtedly the method preferred by the letter-writing and newspaper-reading community of these towns. It was also the preference of the Post Office, although they would doubtless have been more enthusiastic about the arrangements had they not felt that £285 *p.a.* to the T.V.R. was paying well over the odds.

The arrangement remained in force for a couple more years and relations seem to have been reasonably amicable. In April 1848, because the mail coach from Gloucester was often late in arriving at Cardiff, the T.V.R. delayed the starting time of the up train in the morning by half an hour, but undertook to speed it up so that the arrival time at Merthyr was the same.

But then in 1850 there was a major disagreement between the Post Office and the Taff Vale Railway. On 18 June the South Wales Railway opened between Chepstow and Swansea and from 27 July it started to carry the London mail. The mail was sent down to Bristol by train, across the Old Passage to Chepstow, and then on to Swansea by train. It arrived in Cardiff at 5.45 a.m. The same route was followed in the opposite direction, with the up mail train leaving Cardiff at 7.07 p.m. At the same time the T.V.R. made some fairly minor alterations to its train times to ensure convenient connections with the South Wales Railway, and these were agreed to by the Post Office on 26 July. The most significant change was that the last train down from Merthyr was brought forward to 4.30 p.m. from 5.40 p.m. to ensure that there was adequate time to get the mails across Cardiff and onto the London train. It should be remembered that the T.V.R. station in Cardiff was at Queen Street and its trains did not run into the South Wales Railway station, but continued down to the docks along what is now the Cardiff Bay branch. The overall effect was that the mails from Cardiff now arrived in Merthyr at 10.00 a.m. instead of at 10.30 a.m. and they were despatched in the opposite direction at 4.30 p.m. instead of at 1.10 p.m.

Almost immediately following these improvements a group was formed in Merthyr to press the T.V.R. to carry out further changes to its train times with a view to providing a more convenient service for passengers between Merthyr and Aberdare and Merthyr and Cardiff. What the members of this group appear not to have appreciated was that time was required in Cardiff to carry out a further sort of the incoming south Wales mail before it was distributed. After several weeks of three-sided negotiations, involving the T.V.R., the Post Office and the self-appointed representatives of Merthyr, the train times were restored from 2 September to more or less what they had been before 27 July. This meant that the last train from Merthyr was now too late to connect with the up London mail, and so the Post Office had to go back to sending it out on the afternoon train at 1.40 p.m.

This in turn proved unacceptable in Merthyr and yet another petition was presented to the Postmaster General requesting an alteration in the arrangements. Again the Post Office raised the matter with the railway company, but the T.V.R. seems to have lost patience and was not prepared to change the timetable yet again, so the Post Office had no choice but to make alternative arrangements. Rideout had already made provisional arrangements for a mail cart to replace the trains if needs be. Colonel Maberly, the Secretary to the Post Office, thoroughly approved and told the Postmaster General that:

> I have informed Mr. Rideout that he has been quite right
> in making a temporary arrangement for the conveyance
> of the Bags between Cardiff and Merthyr by Mail Cart, if
> necessary, on and from the 2nd September and I have
> directed him to send a further report on this subject as
> soon as possible.

There was some delay in introducing the mail cart and it did not start to run on 2 September. Presumably the Post Office held back in the hope that it might still be possible to reach agreement with the railway company. However, by 10 September it was clear that the T.V.R. was not prepared to make further changes to the timetable and Maberly reported to the Postmaster General:

> It is evident that the Taff Vale Railway must be
> abandoned, as the hours of the trains which the Directors
> propose will give far less accommodation than a Mail
> Cart. I submit therefore ... that a Mail Cart be established
> fitted closely to the Mail Trains at Cardiff, and that Pont-

y-pridd may be made a Post Town as proposed by Mr.
Rideout, the salary being increased from £25 to £35 a
year. Your Lordships will perceive that the private
Pouches now carried by the Railway must cease under
this alteration, but I submit the convenience of the parties
having these Pouches must be sacrificed as they are not
in my opinion entitled to consideration in comparison
with the important town of Merthyr Tydvil. Aberdare will
be required to be served by a Foot Messenger from
Merthyr at wages of £31:5:8 a year and the allowance of
£12:3:4 a year paid to the Aberdare Receiver for meeting
the train may cease. The Merthyr and Station Messenger
can also cease and his allowance of £5:4:3 a year can be
paid to a Messenger who will be required to meet the
Mail Cart at Pontypridd ...

The cost of the new mail cart was £220. Taking all the changes into
account, the overall saving was £17 7s 8d *p.a.* The private pouches to
which Maberly refers were the private bags which were made up for
the ironmasters.[1] Presumably this service had to be abandoned because,
if the mail cart was to arrive at Merthyr at an acceptable time, sorting at
Cardiff had to be reduced to a minimum.

The Assistant Secretary to the Post Office, J. Tilly, reported on the
upshot of the discussions to Sir John Guest, and by the end of
September the arrangements for the mail cart were in place.[2] When it
become common knowledge in Merthyr that their mail would no longer
reach them by train the inhabitants were not impressed. A highly
sarcastic report, obviously submitted by someone living in the town,
appeared in the press.

POSTAL DISARRANGEMENTS. – The Taff Vale
Railway Company having refused (and rightly, too), to
convey the London mails at the price offered by the Post
Office (200*l.* per annum), it has been arranged for a mail
cart to run between Cardiff and Merthyr, as it did

---

[1] cf *The Times* 21 September 1850
[2] A full account of the sequence of events leading up to this development appeared in
the *Cardiff & Merthyr Guardian* 28 September 1850

formerly. Putting a rattle-trap of a cart, with one horse, in opposition to the train? This is about the best joke we have heard of, and redounds to the sageness and ability of the Postmaster General, who, we suppose, thinking to be economical, loses sight of the inconvenience suffered by the inhabitants. Would it not be wise to propose a vote of thanks to him, for the great regard shown for the interests of trade here?[3]

It was indeed ludicrous that the Post Office should be reduced to running a one-horse mail cart between two of the most important towns in Wales. However, the principal concern of the Post Office was to maintain a reasonable service to Merthyr, and if the train times were such that they were unable to do so by rail, then they had no choice but to resort to other expedients. The writer of this piece was mistaken in attributing the re-introduction of the mail cart to penny-pinching on the part of the Post Office. It is true that the Post Office chose to use a mail cart between 1841 and 1846 for reasons of economy, because they were not prepared to meet the payment required by the railway company, but in 1850 the official Minutes make it clear that the *casus belli* was the times of the trains, not the cost.

The mail cart from Cardiff to Merthyr started to run on Sunday 6 October. It was due into Merthyr at 9.00 a.m., so it clearly left Cardiff very soon after the arrival of the down mail train. In the opposite direction the departure from Merthyr was at about 3.30 p.m., just in time to connect with the up London mail at Cardiff. These times compared very favourably with the times that were achieved in August when the mail was still being sent by train. The mail cart must have managed an average speed of about 8 m.p.h. At the same time mail for Aberdare reverted to delivery from Merthyr by a foot messenger, John Chapman, whose appointment was confirmed on 18 October.

The mail cart may have been perfectly adequate for the purpose for which it was intended, but its appearance gave rise to sarcastic merriment among the wags of Merthyr. It was described, no doubt by the author of the previous report as

---

[3] *The Cambrian* 27 September 1850

> ... a structure made up of a few boards nailed together,
> and coated with common red paint, the whole being
> under the care of an eccentric individual, perfectly aware
> of the high office he holds as "Queen's officer".[4]

A further report in the same vein appeared a few weeks later:

> ... we pledge our honour that one of the greatest
> curiosities to be found out of a caravan is the mail cart in
> which the Merthyr letters are brought from Cardiff, in
> consequence of a disagreement between the post-office
> authorities and the railway company. The general
> impression among us is that the company might have
> been more compliant, considering the revenue they
> derive from the town of Merthyr ... truly when we look
> on this car, made up of a few boards daubed with dirty
> red paint, and contemplate a half-starved horse, and an
> illiterate driver, we feel inclined to shout "Chaos is come
> again".[5]

These arrangements were to meet the needs of Merthyr for the next
four years, but Aberdare was now too important to be served merely by
a foot post across the hills from Merthyr. In November 1850 it was
decided that Aberdare should be placed under the new post town of
Pontypridd instead of being under Merthyr and that it should have its
own delivery from Pontypridd. Tenders were invited and the lowest one
for a mail cart from Pontypridd to Aberdare was £80 *p.a.* This cart
probably started in December 1850 or soon after. It was certainly in
existence by March 1851 and in July 1851 was extended to Hirwaun
and Glynneath after the Swansea to Merthyr mail coach was
withdrawn.

Eventually, to the relief, and, it appears, to the surprise of all in
Merthyr, agreement was eventually reached between the T.V.R. and the
Post Office. From 5 July 1854 the London mail, including letters from
London, the west of England and south Wales, started once again to
arrive by rail:

---

[4] *The Cambrian* 11 October 1850
[5] *Cardiff & Merthyr Guardian* 23 November 1850

> On Wednesday morning last, several worshipful citizens
> were startled out of their propriety by an unusually early
> visit from the postman; and much to their surprise and
> gratification they found that the General Post-office had
> adopted the sensible proceeding of sending the Merthyr
> letters by the Taff Vale Railway. Bankers will no longer
> tremble for the safety of their parcels; and less important
> persons will be glad to learn of the new arrangements,
> which not only brings their letters earlier, but gives them
> an additional hour to reply.[6]

The benefits were felt immediately. The mail arrived an hour earlier
than before at 6.15 a.m. and the house-to-house delivery started at 7.15
a.m. In the evening the latest posting time was 5.40 p.m., giving a
whole day to reply to the morning post.

---

[6] *Cardiff & Merthyr Guardian* 7 July 1854

Chapter 6

Dowlais sub-office

Today Dowlais is regarded simply as part of Merthyr Tydfil, but historically it was one of the distinct districts that together made up the multinuclear community that was Merthyr.[1] Deriving from this separate identity, it also has a postal history that until the 1850s is not simply that of a sub-office under Merthyr.

Perched on top of Gwernllwyn Hill, several hundred feet above the valley floor and about a mile from the original village, Dowlais was a company settlement par excellence. It was created by and depended on the Dowlais Iron Company. The Guest family who controlled the company also controlled the settlement, and signs of their influence can still be seen in Dowlais even though the family and the ironworks are long gone.

The first member of the Guest family to be associated with the works was John Guest who was brought in from Shropshire to manage the concern in 1767. In due course John Guest became a partner and on his death in 1787 he was succeeded in the management of the business by his son, Thomas Guest, and his son-in-law, William Taitt. Taitt took control of overall policy and financial management and based himself at Cardiff where he was also responsible for sales. Thomas Guest and after him Robert Thompson undertook the management of the works at Dowlais. Thomas's son, Josiah John Guest, started to take on a managerial role as a young man in 1802, although Taitt seems to have had no great confidence in him: he described him as 'inattentive' and described his writing as deplorable. In 1807 young Guest inherited his father's share in the company and on the death of Taitt in 1815 he inherited his uncle's share as well. He was now the majority

---

[1] For economic and social conditions in Dowlais, see Williams 1988, Evans 1993, Strange 2005, also (with an element of caution) Owen 1972; for Taitt, see also Havill 1983

shareholder in the company. This gave him a power base from which he was able to exercise an influence over the affairs of Dowlais, Merthyr and Glamorgan. His position became even stronger after his election as Merthyr's first M.P. in 1832. Sir John, as he became in 1838, was to remain a constant presence in the affairs of the town and of the county until his death in 1852.

Sir Josiah John Guest (1785-1852)

Dowlais ironworks were established in 1759. The company was slow to take advantage of new techniques and was still producing cast iron alone at a time when the other works had converted to wrought iron. In 1796 its output was considerably lower than that of Cyfarthfa or Penydarren and only slightly more than that of Plymouth (which had only one furnace compared to the three at Dowlais). However, under Taitt, Cort's puddling process was adopted and the manufacture of wrought iron commenced. After this the expansion of Dowlais was rapid, especially because of its successful and early entry in the 1820s into the supply of wrought-iron rails to the railway companies of the world. This success can be attributed initially to Taitt, who turned round a small, inefficient and old-fashioned company, and following him to the organisational and financial skills of Sir John Guest.

The replacement of wrought-iron by steel, the exhaustion of the local ironstone reserves, and higher costs resulted in the decline of the south Wales iron industry from the 1870s onwards. Dowlais was one of the last works to survive; it continued to make steel until 1930, when its closure caused great distress in Dowlais. A subsidiary plant, the Ivor Works, was set up in 1839 and even after the end of steel-making, a foundry on this site continued to work until 1987. It was the final working link with the iron industry which had called Merthyr and Dowlais into existence over 200 years previously.

Under the paternalistic control of the Guests Dowlais had for many years all the characteristics of a company town. Much of the housing stock (about one third) was provided by the company. Josiah John Guest built the parish church, St John's, in 1827 (note the dedication!) and also contributed generously towards the construction of nonconformist places of worship. A handsome stable block (still standing and the without doubt the finest building in the whole of Merthyr and Dowlais) was built in 1820 to accommodate the 500 horses employed in the works. In 1828 a school, regarded as one of the best of its kind, was opened on the upper floor of this building followed by a mechanics' institute in 1829. The truck system was strong in Dowlais, but given the comparative isolation of the place it could be argued this was a necessary provision for the benefit of the employees as much as an opportunity for their exploitation. Indeed, Taitt claimed that prices were lower in the company shop than down the hill in Merthyr and that the operation was non-profitable. Nevertheless in 1807 the prospect of some of the workmen setting up private shops was a cause of concern to Taitt. Guest finally gave up the truck shop in 1832 following the bitter unrest of the previous year.

Compared to the radical community that grew up around the old village of Merthyr, Dowlais was rather more deferential and respectable. Chapel culture was strong and when the temperance movement began in 1837 it was with a mass demonstration organised by the chapels of Dowlais. Slightly earlier, in order to avoid becoming unwilling participants in the riots of 1831, many hundreds of Dowlais men fled the town and sought refuge in Brecon. This difference is reflected too in the differences in personality and in politics between John Guest and William Crawshay of Cyfarthfa. Crawshay, a radical and never quite a gentleman, was altogether a different character from Guest, earnest, socially conformist, and paternalistic. They had little in common and little liking for one another – 'that paltry fellow' as Crawshay referred to Guest on one occasion. But Guest enjoyed great respect, even a sort of popularity, among his workmen, which was enhanced by his marriage in 1833 to the cultivated Lady Charlotte, best known perhaps for her translation of the *Mabinogion*.

The private receiving house

Until 1787, as has been described in chapter 4, letters for Merthyr and Dowlais were normally routed through Brecon where they were collected once or twice a week by a private postman. The route taken by this person was probably across the Beacons on the old Roman road through Pontsticill and then down to Merthyr village through Dowlais. Letters and newspapers for Dowlais were dropped off *en route*.

This arrangement lasted until 1787 when Dowlais, like the other iron companies, started to use Cardiff as their post town. The normal means by which mail reached Merthyr was now by coach to Cardiff and then by a private post messenger, and this applied to Dowlais as well. The Dowlais Iron Company, like the other companies, had its own private bag and this was sent down by a servant from Dowlais to 'the village' (i.e. Merthyr) on post days (Monday, Wednesday, Friday) and handed over to 'the boy', the term that Taitt normally used to refer to the postman in his frequent letters. After the private post had been replaced by an official post in 1804 Dowlais ceased to use a private bag. Rather self-righteously, Taitt told Thomas Guest on 8 July 1804:

> The other works pay Mr. Bird [the postmaster of Cardiff]
> for making up a bag for them – we want no such thing,
> therefore our letters go as others do, in the Common bag,
> sealed up.

In 1804 an official horse post was established between Cardiff and Merthyr and this was upgraded to a mail coach in 1821. Both carried mail for Dowlais as far as Merthyr but from there onwards it was still a matter of Dowlais having to make its own arrangements to get mail to and from the post office in Merthyr. In 1824 a mail coach was started between Merthyr and Abergavenny to carry the London and North mails. The Dowlais London House letter books suggest that Dowlais took advantage of this to secure a faster and more reliable service, since until 1843 letters are frequently addressed to 'Dowlais Works, Abergavenny'. This seems to indicate that Dowlais, like other works along the route of the mail coach, had chosen to have their mail sent to Abergavenny by the General Post and from there carried as a private package by the coach and dropped off as it passed by.

There was a private receiving house at Dowlais before the official sub-office was established in 1843. When the official sub-office was set up the person running it asked for compensation from the Post Office for the loss of income he would suffer as a result. It goes without saying that his request was rejected. It is quite possible that the keeper of this private receiving house also acted as the messenger who collected mail from the post office in Merthyr, but there is no record of this.

Official post 1843

The first proposal for an official receiving house in Dowlais was made in 1823 when Wyndham Lewis memorialised the Post Office to that effect. Lewis was the M.P. for Cardiff and a partner in the Dowlais Iron Company. Rideout reported that the number of letters did not justify the cost of an official post, and that the parties concerned were decidedly against the idea of a Penny Post from Merthyr. It is easy to see why. Under a Penny Post the cost of every letter to or from Dowlais would be increased by 1d. A private messenger between Dowlais and Merthyr every day could carry out the duties for less cost and no less efficiently. No further action was taken in view of this report and the unofficial receiving house and messenger continued in existence for another twenty years.

The question of a sub-office at Dowlais was not raised again until 1843. In January an application was made, presumably by the Dowlais Iron Company and the local tradesmen, for the establishment of an official post at Dowlais. The Post Office was not prepared to act unless the inhabitants were prepared to guarantee that they would meet the expenses. This they were unwilling to do, but they re-submitted their application in March. It was again turned down. However, on 5 August 1843 the decision was taken to establish an official post from Merthyr to Dowlais with no suggestion of a local guarantee. The Surveyor had already been instructed on 3 August to select a suitable person to be the receiver. The expenses were put at £8 *p.a.* for the Receiver's salary and 12s a week (or £31 4s *p.a.*) for a messenger. That nothing was said about the inhabitants meeting the expenses suggests that since April the claims of Dowlais had been put persuasively, very probably by Sir John Guest.

The messenger was expected to collect the letters from Merthyr twice daily and provide a free delivery throughout Dowlais and Penydarren, although at first only one trip a day seems to have been managed. The actual pattern was that the messenger left Dowlais at 5.00 a.m. with letters for the Western mail, due away from Merthyr at 6.45 a.m., for the North and London mails (11.00 a.m.), and for the Swansea mail (1.00 p.m.). Having handed over these letters at about 6.00 a.m. the messenger then had nothing more to do until after the arrival of the Western mail in the evening (4.30 p.m.) when he returned to Dowlais and delivered these letters together with those that had arrived earlier in the day from Swansea (10.30 a.m.) and London and the North (1.00 p.m.). Altogether there were about 500 letters a week for Dowlais at this time. The effect of the establishment of the official sub-office can be seen in the Dowlais London House letters: from August 1843 the inclusion of 'Abergavenny' in the address becomes infrequent, although it never completely vanishes. With the new official post in place, clearly mail was now to be sent straight through to Merthyr on the coach.

A handstamp was issued to Dowlais on 30 August and the first Receiver was appointed, H.P. Powell, whose appointment was confirmed on 6 October. It seems very likely that he had been the keeper of the superseded private receiving house[2] whose claim for compensation had been submitted and turned down on 28 September. However, difficulties were experienced in making a suitable appointment as messenger. On 6 October there was still no messenger but the Postmaster General noted: 'I expect a recommendation every

---

[2] *The Cambrian* 5 November 1852 records the marriage of Mary Ann Adney, 'Stepdaughter of Mr H.P. Powell, for many years grocer and postmaster, Dowlais'. Since his few weeks as official receiver could hardly have counted as 'many years', the implication is that he ran the previous private receiving house

day'. The first appointment was not a success. J. Matthews was appointed but by 26 October Rideout had decided that he was incompetent and he was dismissed. The first Receiver lasted no longer. He had resigned by 26 October. The reason is not recorded.

The new Receiver was Thomas Harding, a grocer, and presumably Rideout also managed to find a new messenger, but it was not long before there were further difficulties. In April 1844 an anonymous complaint was made against the Receiver. Maberly reported that

> As the signature to the enclosed complaint against the
> Receiver of Letters at Dowlais appears to be fictitious, I
> submit that it is unnecessary to take any further action in
> the matter.

However, the complaint was probably not just a matter of petty malice, for less than a year later, on 15 March 1845, Rideout reported that he could no longer place any confidence in the Receiver 'after what has occurred', and he was dismissed. The Minutes record neither the nature of the original complaint nor the conduct of the Receiver that led to his dismissal.

In October 1845 a general request was put forward from Dowlais that the messenger be re-timed so as to connect with the London mail rather than wait for the arrival of the Western mail. In other words, the messenger would leave Merthyr at about 1.00 p.m. after the arrival of the London mail, and letters coming in later in the day by the Western mail would have to wait until the following morning. The petitioners were so keen on this change that they were willing to see delays to letters from Cardiff and Bristol in order to enhance the London service. This is entirely understandable: in the 1840s Dowlais was at the peak of its success and claimed to be the largest ironworks in the world. They were manufacturing rails for the railways of the world and the correspondence between their London office, where the orders were generally taken, and the works was immense. The change was agreed, and additionally it seems that the second despatch to Dowlais promised in 1843 was initiated either at the same time or soon afterwards. By 1848 there were two despatches from Merthyr to Dowlais at 11.00 a.m. and 2.30 p.m. which corresponded to the arrival of the Western mail by train in the morning and the London and North mails in the afternoon. Similarly there were two despatches from Dowlais to Merthyr, 9.00

a.m. for the Western and Swansea mails and 1.00 p.m. for the London and North mails.

Following the dismissal of Harding in March 1845 there was a lengthy delay until a new Receiver, Abraham Wyse, was appointed in October 1845. He died less than a year later in July 1846. The failure of his superior, Rhys Davies, the postmaster of Merthyr, to report this fact led to his receiving a sharp reprimand from the Post Office. Wyse was replaced in September by Frederick Evans, the fourth receiver in three years.

Sub-office status

In 1847 Sir John Guest made further efforts to get the service to Dowlais improved. The mail coaches to and from Abergavenny, carrying the important London mail, passed through the town *en route* to Merthyr but did not put off any mail. Instead Dowlais had to wait for the mail to be taken off the coach and sorted at Merthyr, and then for the Dowlais mail to be brought back on foot. Similarly, in the opposite direction, outgoing mail had to be taken down to Merthyr by foot at 11.00 a.m.; it then passed back through Dowlais again three hours later on its way to Abergavenny. Guest requested that Dowlais should be up-graded from receiving house to a sub-office dependent on both Abergavenny and Merthyr – dependent on Abergavenny so that mail from London and the North could go directly to Dowlais; dependent on Merthyr so that the Western mail could continue to be forwarded to Dowlais after the arrival of the train from Cardiff. This proposal would have required a separate bag for Dowlais being to be made up at Abergavenny, and Rideout reckoned that this would require an additional clerk there, which seems rather extreme. There would also be additional costs at Dowlais if it were made a sub-office: they would increase from £39 5s 8d *p.a.* to £59 6s 5d *p.a.* But there was certainly some merit in Guest's case. The number of letters had increased from 500 a week in 1843 to 1,100 a week in 1847. Rideout reported on the situation in March 1847 but, not surprisingly, on 2 July the Treasury refused permission for the second clerk at Abergavenny. Rideout was then instructed to see whether anything else could be done to improve the service to Dowlais, but in September he reported again that the only way an improved service could be offered was if the second clerk was employed.

Guest was duly informed and there the matter rested for a time, but in March 1849 the decision was made to upgrade Dowlais to sub-office status, and on 1 April it was made a Money Order Office, earlier than either Aberdare or Pontypridd. In October 1849 a direct bag was established between Abergavenny and Dowlais. The wages of the Dowlais letter-carrier were reduced from 12s a week to 10s on the grounds that he would no longer have to make the journey to and from Merthyr, only carry out the delivery in Dowlais and Penydarren. The saving of £5 4s 3d *p.a.* would go towards the £25 needed to pay for extra assistance at Abergavenny. This works out at rather less than 10s per week, so clearly the additional clerk was not full-time. The effect of the new arrangement was that the London and North mails now arrived at Dowlais two hours earlier and were despatched an hour and a half later, which gave 3½ hours extra to reply to the mail, a significant improvement. The Western mail from Bristol, Cardiff and all parts of south Wales was presumably sent up to Merthyr in the usual way and then transferred to the Abergavenny coach.

The new arrangements were not introduced immediately. In April 1850 a note appears in the Post Office Minutes recording the discontinuance of the post from Merthyr to Dowlais 'for the Postmaster General's information', and this must mark the transfer of the Dowlais mail to the coach from Abergavenny.

The new arrangements did not last long. In January 1852 Rideout recommended that Dowlais be served from Merthyr rather than from Abergavenny. Now that the London mail was being carried into Wales by rail, the arrival at Cardiff was so much earlier that there were no benefits in routing the London mail to Dowlais through Abergavenny; however, the North mail continued to be sent by coach to Abergavenny for a few more years. The services of the additional clerk at Abergavenny were dispensed with and the messenger from Merthyr to Dowlais was reinstated, this time at a cost of 8s per week. Letters from the London House to Dowlais works were now generally addressed to Merthyr.

This duties of this messenger were described by Trollope in July 1853 as part of his review of the local posts in south Wales. By this time the messenger's wife had been given the position of assistant letter-carrier at Dowlais. The walk was undertaken twice a day on weekdays and once on Sundays. The route was given as Penydarren, Twenty Houses, Penwernfawr, Duckspool, Penywern, Gwillwynfach, Pengarnddu. The exact location of Twenty Houses and Duckspool are no longer known; Gwillwynfach must be Trollope's attempt at Gwernllwynfach. The messenger himself was to deliver at Penydarren, Main Street of Dowlais, Twenty Houses, Penwernfawr and Duckspool; he was also to collect the outgoing bag every evening from Dowlais office. His wife, the assistant letter-carrier, was responsible for 'that part of Dowlais to the left of the Main Street, also Penywern, Gwillwynfach & Pengarnddu'. She was also to deliver the North mail which came in on the coach from Abergavenny – or as it was now rather more prosaically known, the 'omnibus'. A separate Dowlais bag was still being made up at Abergavenny for the North mail and local letters and presumably this continued to be the case for as long as the coach continued to run.

Meanwhile the personnel difficulties which had affected Dowlais ever since the establishment of the post office were continuing. Between 1846 and 1850 three more receivers came and went. Added to this the letter-carrier, Thomas Thomas, was arrested in August 1849 for secreting letters, which must mean that he extracted those that he thought might have included money. His departure was unlamented:

> It should be recollected in the case of the late Dowlais
> Postman that the lower wages he received was doubtless
> the incentive to his bad career, now happily for himself
> and all terminated[3]

---

[3] *The Cambrian* 8 March 1850

Official Receivers and Sub-Postmasters at Dowlais

	Appointed	In office	Resigned
H.P. Powell	1843		1843
Thomas Harding	1843		1845
Abraham Wyse	1845		1846 (died)
Frederick Evans	1846		1847
John Peters	1847		1849
Joseph Jones	1849		1850
?	1850		
Thomas Hopkins		1858	

Chapter 7

Aberdare and the Cynon valley

The development of Aberdare in the Cynon valley has many parallels to that of Merthyr although in every respect on a reduced scale. Like Merthyr, Aberdare was the centre of one of the ancient upland parishes of Glamorgan and its initial growth was due to the ironworks which were established in or near the town.

The first ironworks, at Hirwaun, was set up in 1757, a couple of years before Dowlais. It was probably the first coke-fired furnace in south Wales. The other three ironworks were all established well after the main works at Merthyr had been set up. These were the Llwydcoed (or Aberdare) ironworks (1800), the Abernant ironworks (1801), the Gadlys works (1827) and finally Crawshay Bailey's Aberaman ironworks (1845). The existence of the ironworks led to an improvement in transport facilities: a turnpike road from Aberdare to Navigation (as Abercynon was then known) was started in 1803 and completed in 1810, and the Aberdare Canal, a branch of the Glamorganshire Canal, was opened in 1812.

The Aberdare iron industry was never as successful as that of Merthyr, which can in part be attributed to the fact that Aberdare coal did not coke particularly well, unlike the coal from around Merthyr. In the eleven years 1829-39 the most successful of the Aberdare concerns, the Aberdare Iron Company, exported 40,000 tons of iron less than Penydarren, the least productive of the Merthyr works (Aberdare: 96,396 tons. Penydarren: 135,584 tons) and in 1839 Aberdare and Hirwaun had only eleven furnaces in blast compared to over thirty in Merthyr. But the population figures show that the Aberdare iron industry was far from being a complete failure, for with the exception of the post-war depression decade 1811-21, the population expanded at an average rate approaching 10 *per cent p.a.* Even so it constantly lagged behind that of Merthyr, both absolutely and in terms of the overall percentage growth rate.

The failure of Aberdare to develop an iron industry as successful as that of Merthyr is also reflected in its slower growth into a self-contained urban centre with a full range of facilities. In 1835 Aberdare, with a population of perhaps around 5,000, is recorded as having had only four grocers, one tailor and one chemist compared to 77 grocers, nine tailors and six chemists in Merthyr and Dowlais. The population of Merthyr (including Dowlais) was, of course, much greater (around 25,000) but the discrepancy is still marked. Aberdare had one grocer to 1,250 head of population; Merthyr and Dowlais had one grocer to 325 head of population.[1] The same is true of professional services: as late as 1852 there were still no banks or solicitors in Aberdare.

The decade in which the most rapid growth occurred was the 1840s, when the population increased by 132 *per* cent. This was the decade when the economy of the Cynon valley moved from dependence on the iron industry to the mining of coal as a sale commodity rather than as a raw material for the blast furnaces. Aberdare was now no longer a rather unsuccessful rival to Merthyr in the iron trade but the most productive centre of the sale coal industry in south Wales, a position it held until overtaken by the Rhondda in 1884. Between 1837 and 1853 a total of sixteen deep pits were sunk in the valley in and around Aberdare.

~~~~~~

Early postal service

Until the 1830s there is little evidence for the nature of Aberdare's postal service. In about 1790 Woodcock recorded a delivery from Cardiff to Aberdare, but there is no indication as to how the delivery was made or who employed the messenger. Clearly it must have been a private delivery and presumably it worked in with the messenger from Cardiff to Merthyr, meeting him at some point en route. As Freeling remarked in 1834, when an official post was established, ' ... the inhabitants of this village have hitherto received their letters by private arrangement'.

In 1801 a postman by the name of Morgan yr Allt was recorded as carrying 'the few letters for the Aberdare Valley'.[2] In 1815 (probably)

---

[1] *Pigot's Directory* 1835.
[2] 'Some notes on the early postal service at Aberdare' 1973

his place was taken by William Dyke. By 1831 the *Monmouthshire Merlin* calculated that Dyke had walked over 126,000 miles in the course of his duties:

> PEDESTRIANISM EXTRAORDINARY. – Mr. William
> Dyke, the Aberdare postman, has travelled the following
> miles:- Walked from Newbridge to Aberdare and back,
> 5½ years, computing the various places he has to call at,
> cannot be less than 27 miles a day; six days a week will
> make a distance of 46,332 miles. Walked from the
> Navigation to Aberdare and back, 10½ years, 21 miles a
> day, seven days a week, will be 80262 miles, in all
> 126,594 miles in 16 years. Mr. Dyke is now 62 years old
> and continues to perform the business with ease to
> himself and satisfaction to the public.[3]

These figures suggest that Dyke started his job towards the end of 1815[4] and that until April 1821 he met the Cardiff to Merthyr horse post at Newbridge (or Pontypridd, to use the modern name). He allowed himself one day off a week although the Merthyr horse post operated every day, and had done so since 1812. With the introduction of the Merthyr mail coach in 1821 Dyke's daily walk was reduced since he now met the coach at Navigation (now known as Abercynon). It is not clear why he could not also have met the horse post at Navigation and so saved himself six miles a day. The distances quoted indicate that Dyke cannot have walked much beyond the modern centre of Aberdare and certainly not as far as Hirwaun. This is rather surprising, since Hirwaun was one of the main ironworking centres in the Cynon valley. The mail cart between Neath and Merthyr which started to operate in 1834 served Hirwaun (see chapter 8), but there must have had some means of sending and receiving letters before that date. In 1828 the normal number of letters for Aberdare was said to be ' ... some Three Dozen letters a day'.[5]

---

[3] *Monmouthshire Merlin* 12 March 1831 reproduced in *The Cambrian* 3 May 1851
[4] When Dyke's death was announced in *The Cambrian* 28 February 1835 he was said to have been the Aberdare postman for nearly 30 years; but '30' could be a typographical error for '20'
[5] 'Some notes on the early postal service at Aberdare' 1973

Official post 1834

In 1834 the first official post to Aberdare was established. The campaign that led to the establishment of a direct mail between Neath and Merthyr also resulted in a demand for improved facilities for Aberdare. At the time Freeling described it as ' ... a village 4 miles distant from Merthyr ... ', which reflects the lack of urban development. On 15 January 1834 a meeting was held in Merthyr with John Guest in the chair. After those assembled had expressed their views on the desirability of a direct mail from Neath to Merthyr, they moved on to consider the question of an improved service for Aberdare. Guest, who had met Rideout some three weeks previously, explained that the Post Office was willing either to 'carry on the General Post direct to Aberdare' or to establish a Twopenny Post under Merthyr. Extending the General Post to Aberdare would have resulted in Aberdare becoming a post town, a status it was not to attain until 1855. Quite how this would actually have been implemented was not discussed. If this course were to be adopted, then every letter between Aberdare and Merthyr would carry a charge of 4d, the minimum rate for a General Post letter; on the other hand, letters coming in from outside would cost no more than they did already, since the difference in mileage would be negligible. If a Twopenny Post were to be set up, then the cost of a letter between Merthyr and Aberdare would be only 2d, but all letters from outside would carry this charge in addition to the General Post rate. The meeting preferred the first option as being more advantageous and moved that Guest should apply for the extension of the General Post to Aberdare.[6]

However, in the end the Post Office preferred the other option and a Penny Post between Merthyr and Aberdare commenced on 21 February 1834. The first Receiver was John Jones, a druggist, who was sworn in on 22 February.[7] His premises were at 1 Commercial Place (now Victoria Square). Mail was routed through Merthyr and a messenger was appointed to carry the letters on to Aberdare on foot. This was probably William Dyke again, since when he died on 18 February 1835 the announcement in the press described him as 'the Aberdare postman

---

[6] *Glamorgan, Monmouth, and Brecon Gazette, and Merthyr Guardian* 18 January 1834
[7] *Aberdare Leader* 8 December 1934. I am indebted to Mr Gerald Richards for this and several other references used in this chapter

for nearly 30 years' with nothing to suggest that he was no longer in post. He must have taken the road over Aberdare Mountain and his day started early, since he was due to arrive at Merthyr at 6.45 a.m. with the outward letters from Aberdare so as to connect with the coaches to Cardiff and Abergavenny which both left at 7.45 a.m. On his return journey he left on the arrival of the Abergavenny (London) and Cardiff mails at 6.30 p.m. During the day he was presumably at leisure to amuse himself in Merthyr as he chose or to walk back home to Aberdare before returning to Merthyr once again in the evening. His load may not have been heavy, but it was a long day. The pay, 2s a day, 14s a week, was the same as the town postman in Merthyr received at this time for a day's work that was shorter both in terms of hours and of the distance to be covered.

The cost of the messenger was estimated at £36 10s *p.a.* with a further £4 for the Receiver. The income that could be expected from the Penny Post was put at £33 4s *p.a.* This suggested that the service would be operated at a loss, but Freeling noted that with Merthyr now the post town an increase in General Post receipts could be expected because of the greater distance that the letters would be carried by the Post Office. Even so, he recommended that the service be made experimental for one year in the first place. The anticipated income translates into 7,968 letters, or roughly 22 a day, which is rather less than the three dozen a day which Dyke was supposed to have carried in 1828. Freeling's estimate turned out to be very accurate: in the year ended 5 July 1836 the gross revenue of the post was £33 15s 2d.

Taff Vale Railway

Mail continued to be delivered to Aberdare by foot from Merthyr until 1848. The times of despatch and arrival changed slightly as the times at Merthyr changed, but the basic pattern of despatch early in the morning and arrival in the evening remained unchanged. This, of course, made a same-day reply impossible and an overnight reply inconvenient. In the summer of 1846 mail started to be carried to Merthyr by the Taff Vale Railway which resulted in a much more congenial regime for the Aberdare messenger: the outward mail was now despatched at 8.30 a.m. with the incoming mail brought back at 11 a.m. There was no longer a second trip to Merthyr in the evening. The names of two of these messengers have been preserved – Richard Morgan, who resigned in January or February 1846, and his replacement, Thomas Howell.

The Aberdare Railway (a subsidiary of the T.V.R.) was opened in April 1846 and in February or March 1848 mail started to be sent to Aberdare by train and sent directly rather than via Merthyr. The foot messenger was dispensed with, but the Receiver at Aberdare had to collect the bags from the station, for which he was paid 8d a day. The overall increase in the cost of serving Aberdare in this way was £10 13s 4d *p.a.*, but the Post Office accepted it because it would allow a reply to a letter to be sent on the same day as it was received – so long as the writer was snappy about it. The mail arrived at 10.30 a.m. and was despatched at 1.00 p.m. which gave Aberdare the same level of service as Merthyr. Because the messenger from Merthyr had been discontinued it was no longer possible for Aberdare's London and North mails to be sent on the coach from Abergavenny. They were therefore routed through Cardiff in the same way as the Western mail. In any case this provided a better service than if they had still been routed via Merthyr, since the coach from Abergavenny did not reach Merthyr until 10.30 a.m.

From July 1850, following the opening of the South Wales Railway, mail from London to south Wales was routed via Bristol rather than Gloucester, and travelled by rail all the way except across the Severn between Bristol and Chepstow. However, in October a dispute between the Post Office and the T.V.R. resulted in the re-instatement of a mail cart between Cardiff and Merthyr and consequently of the foot messenger across the mountain to Aberdare (see chapter 5). This situation did not last long. In November 1850 plans were made to transfer responsibility for Aberdare from Merthyr to the newly established post town of Pontypridd and to serve Aberdare by means of a mail cart from Pontypridd. Presumably mail was to be carried between Cardiff and Pontypridd by the Merthyr mail cart. Pontypridd took control on 10 February 1851, which may well be the date on which the mail cart started to run: it was certainly in service by March 1851. The mail cart remained Aberdare's main link with the outside world until 1854, when the Post Office's differences with the T.V.R. were settled and the mail was once again carried to Aberdare and Merthyr by rail. In 1852 the mail cart arrived at Aberdare in time for a delivery at 11.30 a.m., with a return at 2.00 p.m., which was as good a service as had been offered when the T.V.R. carried the mails.

Local arrangements

Within Aberdare arrangements for a free local delivery were put in hand in May 1845 with instructions that the messenger from Merthyr should also carry out a house-to-house delivery on his arrival back from Merthyr with the letters. There is no record of any additional payment being authorised. However, the service did not commence and four months later it was discovered that the messenger was unable to carry out the task because he was illiterate. He had been appointed by the postmaster of Merthyr, who was responsible for Aberdare, with no reference to the local Surveyor or to the General Post Office establishment, and presumably the postmaster was not prepared to admit that he had made a bad appointment. Orders were given that the illiterate messenger should be dismissed. In November 1850 arrange-ments for a free delivery at Aberdare were once again under discussion, this time at a cost of £7 16s 5d *p.a.* The amount allowed was the equivalent of 3s a week, which must be an indication of how much lighter the duties in Aberdare were compared to Merthyr, where the original letter-carrier received 14s a week, and the second one, appointed in 1850, 6s. However, by December 1853 the letter-carrier who undertook the local delivery was receiving the sum of 18s per week and delivering to an area that included the town of Aberdare itself, the area around the head of the canal, and the ribbon development connected with the Abernant ironworks which stretched north-east along Abernant Road, past the ironworks and up to the newly opened station on the Vale of Neath Railway

It is not known how long John Jones remained as Receiver, but his place was taken in about 1838 by Ann Thomas. Her father had previously been the proprietor of the *Black Lion*, the principal inn in the town, and she may well have run the post office from the inn. Within a year or two Ann married Robert Jones who had taken over the *Black Lion*. As a married woman she was no longer able to hold the office of Receiver and so, nominally at least, the post office was transferred to

The Black Lion Hotel *in Wind Street, Aberdare which served as the town's post office from about 1838 to 1852*

her husband.[8] By May 1845 about 400 letters a week were passing through Aberdare and the Receiver was given an increase in salary from £4 to £6 *p.a.* However, this was ' ... on the distinct understanding that the duties of the office are performed & the letters kept and sorted in a place distinctly separate from the Bar of the Inn'. Jones took his increase, but in September ' ... he still refuses to provide a fit and proper place for the performance of his official duties' and so his dismissal was recommended. In January 1846 it was noted that a vacancy for a Receiver existed at Aberdare., but in February 1846 Robert Jones was re-appointed, presumably because there had been no other suitable applicants. In April 1851 there is a further minute recording the appointment of Robert Jones as Receiver at Aberdare yet again, but this is probably no more than confirmation of his position following the

---

[8] A similar situation arose at Cowbridge in 1838 when the postmistress, Mary Llewellyn, married Thomas Lister, also at Maesteg in 1850 when the receiver, Miss Ballard, married the Revd David Phillips

transfer of control from Merthyr to Pontypridd which had taken place that February.

Whilst Robert Jones was nominally the Receiver of Aberdare, it seems pretty clear that the real work continued to be done by his wife and that whatever the legal position might have been, she was regarded as Postmistress by the townspeople. She resigned from the post in 1852 for reasons which were not made clear.

> ABERDARE POST OFFICE. – We regret to hear that the highly-popular and respected post-mistress, Mrs. Jones, has resigned her trust, which she so honestly and faithfully discharged for a period of fourteen years. The inhabitants of Aberdare heard of her resignation with great regret. We hear that a subscription is on foot to present Mrs. Jones with a tangible proof of the respect in which she is held by her neighbours generally.[9]

The promised presentation was made later in the year and Ann Jones received fulsome praise in the newspaper account of the event:

> To those acquainted with Aberdare, the former post-mistress, Mrs. Jones, of the Black Lion Inn, must be well known to them, not only as a hostess (and the daughter of old Levy Thomas celebrated in former times for his bacon and eggs), but for her great attention and politeness as a post-mistress, which she conducted for 14 years, with little pecuniary remuneration for the services required; the remuneration was trifling comparatively with the duties of a post-office, in a straggling place such as Aberdare is, with a population of near 17,000 inhabitants. Without fee or reward this excellent woman who was always anxious to oblige her neighbours, and all those who had anything to do with the post-office – the gentry, tradesmen, and the inhabitants generally – can amply testify the attention and politeness always received there … [10]

---

[9] *Cardiff & Merthyr Guardian* 10 July 1852
[10] *Cardiff & Merthyr Guardian* 11 December 1852

She was presented with a silver tea-pot, a cream jug and a silver sugar basin by the vicar of Aberdare, the Revd John Griffith, a distinguished figure in the Victorian Church in Wales. Robert Jones responded on behalf of his wife. John Jones the druggist, described as 'now the oldest tradesman in the town' and its first official Receiver, was also present and he 'made some remarks as to the ancient name of Jones which he was proud to bear, and dilated upon the pride and satisfaction he felt upon being present on such an occasion'. One can just imagine it!

Following the resignation of Ann Jones 'Mr Smith', obviously a member of the Surveyor's staff, was sent to take Aberdare in charge on 31 May 1852 and the following day, 1 June, the vacancy at Aberdare was reported to the Treasury. By July a new Receiver had been appointed, William Morris, a printer and stationer.[11]

Until 1851 Aberdare came under Merthyr, but on 10 February 1851 control passed to Pontypridd. Money Order Office status was granted on 10 March 1851. In 1855 Aberdare was made a post town and took control of the Cynon valley sub-offices. A previous request for this status had been rejected in 1849.

Figures exist to show the growth in business at Aberdare over a period of 25 years, although in some cases it is not clear whether the figure is for incoming or outgoing mail only, or whether it is for all items that passed through the office.

Year	Letters per day	
1828	36	
1834	22	
1845	66	(400 per week)
1848	200	(1,398 per week)
1853	252	(7,817 in January 1853)

It is noticeable that the greatest period of growth was in the later 1840s rather than in the earlier part of the decade immediately after the introduction of the uniform penny post. The postal reforms of 1840 certainly led to an increase in traffic but the real increase was caused by

---

[11] In 1861 Morris was accused by a former servant of fathering her illegitimate child. An extensive report of the case appeared in *The Cambrian*, 8 March 1861. Her claims were proved and he was ordered to pay maintenance but retained his office until being declared bankrupt in 1883

the expansion of the steam coal trade and the accompanying growth in population and commercial activity.

Sub offices

Sub-offices started to be opened in the Aberdare district in the 1840s, a further indication of the growth in population and economic activity caused by the success of the coal industry and the opening of the new deep mines.

The first of these was at Hirwaun in 1843. Until 1851 it was served by the Swansea-Merthyr mail coach. The coach ceased in 1851 and was replaced by a horse post between Aberdare and Glynneath. A full account is given in chapter 8.

The next sub-office was at Aberaman, down the valley from Aberdare and the site of an ironworks set up by Crawshay Bailey in 1845 and of a pit sunk by him in the same year. A sub-office was set up in 1847 in response to representations made by the Revd John Jones, who was presumably a local minister and as such a spokesman for the community. On 9 July 1847 he was notified by Charles Johnson of the G.P.O. that

> ... the Post Master General has been pleased to sanction
> the establishment of an official Post from Aberdare to
> serve Mill Street (Trecynon) and Aberaman, which will
> commence as soon as the Lords of the Treasury have been
> pleased to nominate a Receiver at the latter village ... [12]

A handstamp was issued to Aberaman on 24 July,[13] so the receiving house probably opened shortly afterwards. The idea of a single post to serve both Trecynon and Aberaman seems rather bizarre, since the two villages lie on opposite sides of Aberdare. There is no record in the Post Office Archives of the decision to set up the receiving house at Aberaman and the appointment of a receiver or of the arrangements which were to be made to serve it.

In March 1852 the horse post from Aberdare to Glynneath was discontinued after less than a year's operation. By way of replacement the Surveyor recommended that Glynneath mail should be delivered by

---

[12] 'Some notes on the early postal service at Aberdare' 1973
[13] Broomfield 1996, 76

rail from Neath and that Hirwaun should be served by extending the Mill Street (Trecynon) foot post through Penywaun to Hirwaun. A new sub-office was to be established at Mill Street to cover deliveries to Trecynon, Robertstown and Llwydcoed. But if the Trecynon messenger was now to go as far as Hirwaun, clearly he could not cover Aberaman as well, so the sub-postmaster of Aberaman was to have his salary raised from £3 to £4 *p.a.* with an additional £5 for delivery in Aberaman and Blaengwawr. By December 1853, at the time of Trollope's review, the Hirwaun foot post had been further extended to the quarry village of Penderyn 'as far as the Church'.

These proposals were accepted although there seems to have been some delay in implementing them. There is no record of an appointment of a sub-postmaster at Mill Street and the first handstamp was not issued until July 1853. A further enhancement, obviously connected with these revised arrangements, was the creation of a sub-office at Cwmbach, near Aberaman. This was agreed in February 1853 and the first handstamp was issued in April 1853. Deliveries and collection were made the responsibility of the sub-postmaster at Aberaman. Cwmbach was the site of Thomas Wayne's pit of the same name, the first of the great steam coal collieries of the Cynon valley, sunk in 1837.

These arrangements were all in place by December 1853 when Trollope reported on his review of the Pontypridd rural posts.

Mountain Ash

In Mountain Ash, a colliery settlement which grew up to house workers in the pits sunk by Thomas Powell between 1840 and 1850, the first post office was set up in 1852. The decision to establish it was minuted on 28 October and a handstamp was issued in December. The sub-postmaster was allowed a salary of £3 *p.a.* with an additional £2 *p.a.* for delivery. Until 1854 mail was delivered and despatched by means of the mail cart from Pontypridd to Aberdare. When the mail contract reverted to the T.V.R. in 1854, the Aberaman foot post was extended through Abercwmboi to Mountain Ash.

A fine description of the duties of the Mountain Ash postman in the early 1860s survives and it is worth quoting in full, both for the historical information which it contains and because of the feel which it conveys of the milieu within which the Post Office operated at this time. No doubt the situation which it describes would apply equally to any other valley at this time, partly still rural, partly in process of rapid industrialisation.

> In 1861 there was no resident postman in Mountain Ash, although there was a Post Office in the town for posting letters (170) and purchasing postage stamps. The postman travelled down from Aberdare with his bag containing letters, through Aberaman and Abercwmboi, across the field passed [*sic*] the Dyffryn House, residence of Lord Aberdare, and into Mountain Ash. He distributed the few letters he had in the area, and having finished his work he waited there for some time until the afternoon. One such postman worked as a cobbler in his spare time, and when the afternoon post was ready to go out, the cobbler's tools were placed aside and he then changed into the official uniform of Her Majesty; a cobbler in a few minutes being transformed into a postman! Off he would go, towards the Post Office to collect the letters to place in his bag, and off went the post with all the vigour of a Mail Steamer leaving for Dublin! He headed towards Aberdare passing Dyffryn House once more to collect letters. The residents at Dyffryn House would not take their letters to the Post Office because the postman called for them. From Aberdare the letters were sent through the mail to various places, and it was necessary at that time, and more recently also, to address letters to 'Mountain Ash, near Aberdare', otherwise they would end up in the 'Dead Office', having failed to locate the place.

Notes

170. John Griffiths (1805-1891). The first Postmaster of the town; resided at Fair View Villa, Caegarw. The Post Office was situated at Ffrwd Crescent, Caegarw. John Griffiths was one of the first to settle in Mountain Ash from around 1845, at which time he witnessed the growth of the town from a village of about 300 people to a town of 8,000 in 1875. A member of the Local Board of Health for sixteen years and one of the founders and trustees of

Providence Church, Union Street. Also a lay preacher. He retired as Post Master in July 1875 (aged 70) and died 23rd May 1891 aged 86 years.[14]

Receivers and Postmasters of Aberdare

	Appointed	Resigned	
John Jones	1834	c1838	
Ann Thomas	c1838	c1840	(married)
Robert Jones	c1840	1852	
William Morris	1852	1883	(bankrupt)

---

[14] Jones c1990, 69

Chapter 8

The Neath Valley, including Glynneath and Hirwaun

Compared to the valleys that lie to its east, the Neath Valley has remained comparatively unindustrialised. This is because of an underlying geological feature, the 'Neath trough', which follows the line of the valley and has the effect of pushing the coal reserves down to a depth which made exploitation uneconomic. Because the coal was out of reach other industries were not attracted to the area. However, at either end of the valley the coal reserves come close to the surface and this led to industrial activity and population growth, albeit on a more limited scale than in the heart of the coalfield.

The upper end of the valley is at the northern edge of the coalfield where the coal measures outcrop alongside limestone and millstone grit. This range of resources led to the development of a small township at Glynneath from the 1790s onwards following the completion of the Neath Canal. A number of industries were established which reflected the variety of the underlying geology. Coal and ironstone were mined and these were transported down the canal to the Neath Abbey ironworks to the west or overland to the ironworks of Aberdare in the east. In the 1840s, several ironworks were erected in and around Glynneath to exploit the application of the hot-blast method to iron smelting. It was claimed that this method, which had recently been developed at Ynyscedwyn in the neighbouring Swansea valley, would enable the local anthracite to be used to produce iron every bit as good as steel. This was very soon found not to be the case and the local iron industry never developed to any great extent. Other industries were more successful, including coal mining, the quarrying of limestone, the manufacture of gunpowder and, in particular, the manufacture of silica furnace bricks using the silica sand which was to be found in the gritstone formation. These activities were all on a fairly modest scale and Glynneath remained a compact community with none of the ribbon development and overcrowding that characterised many of the other valleys.

Glynneath lay on the floor of the valley. Several hundred feet higher up, and across the watershed, was Hirwaun. Although Hirwaun has always been linked with Aberdare for administrative purposes, and geographically lies at the head of the Cynon valley, it makes more sense for present purposes to consider it in conjunction with Glynneath and the Neath valley: it lay on the main road from Neath to Merthyr and for postal purposes it was served from one or other of these places. This road was authorised by an Act of Parliament of 1795. Work started in 1797 and it was completed in about 1811. Hirwaun was the site of what was probably the first coke-smelting iron furnace in Wales, established in 1757. Iron continued to be the mainstay of the place until it was replaced by coal in the mid nineteenth century. In 1850 its population was about 3,000.

Towards to an official post

The first official postal service in the Neath valley was set up in 1834. There is little evidence as to the arrangements that may have existed before that date. In about 1790 Samuel Woodcock recorded that a bag was being made up at Cardiff for 'Aberpargam', which must be Aberpergwm, near Glynneath, the seat of the Williams family, the principal landowners in the upper valley. Presumably this bag was put off the mail coach at Neath and a family servant collected it. One presumes that there was also some sort of private post connecting Hirwaun to Merthyr. Unfortunately, in compiling his summary of services in south Wales, Woodcock never got as far as the letter N so we do not have the detailed information for Neath that we have for places with names at the beginning of the alphabet.

In 1827 the first application for a mail service between Swansea and Merthyr to be reported in the Post Office's records was made. Freeling was not impressed, since he suspected that the request came from the proprietors of an existing coach running out of Swansea who wanted the mail contract as a way of subsidising their operation. At this time two stage coaches operated between the two towns, the *Imperial* and the *General Picton*: perhaps they were both barely profitable and the owners of one or other had the idea that a mail contract would give them an edge over their competitor. Freeling believed that there was no real public demand for a direct mail service, and that the existing service between Swansea and Merthyr, via Cardiff, was perfectly adequate: a letter posted in Swansea in the evening was delivered in Merthyr the following morning. It was estimated that the cost of the service would be £100 *p.a.* and no further action was taken.

A few years later, in 1831, a further application was made, this time by a Mr Williams, presumably William Williams, the squire of Aberpergwm. He seems to have suggested that the service could be provided economically by means of a mail cart which would also carry passengers. The surveyor pointed out that this would infringe the rule forbidding the conveyance of passengers along with the mail bags in a mail cart, and in any case there were many further objections. What exactly these were was not stated, but probably the poor condition of the road was high among them. By this time the estimated cost of providing the service had risen to between £150 and £170 *p.a.*, and again no further action was taken.

A further campaign was started in 1833 when a coalition of commercial interests in Swansea and Merthyr and the landowners of the Neath Valley came together to press for the establishment of a mail coach from either Swansea or Neath to Merthyr. Swansea interests favoured this route because it was seen as a means of improving the London mail which at that time was routed by Bristol and the New Passage. A more reliable service would be possible, it was claimed, if Swansea's London mail was routed via Gloucester and the Abergavenny-Merthyr coach, which could then be extended to Neath and Swansea, thus securing an earlier arrival and a more reliable service.

A public meeting was convened in Swansea on 28 August 1833. It passed a resolution to petition the Postmaster General to have the mail routed through Abergavenny and Merthyr. A committee was formed to pursue the matter, which included the industrialists and M.P.s, John Henry Vivian of Swansea and Josiah John Guest of Dowlais. In the following weeks meetings were also held at Merthyr and Neath which passed similar resolutions and formed committees to collaborate with the Swansea committee. William Williams of Aberpergwm was among those on the Neath committee.[1]

The Postmaster General, acting of course on the advice of Freeling and the local Surveyor, Charles Rideout, decided against establishing a mail coach between Abergavenny and Swansea. It would have undermined the financial basis of the existing coaches between Merthyr and Cardiff and Merthyr and Abergavenny, and in any case the state of the road made it impossible to contemplate running a mail coach over it. However, a promise was given that some other form of postal communication between Neath and Merthyr would be established. This probably indicates that Guest and Vivian had made fairly forceful representations to the Post Office on behalf of their constituents.

In due course Rideout was sent down to Swansea to make further enquiries into the benefits that might be expected to arise from a direct post to Merthyr, its costs, and the means by which it could be implemented. He arrived on 27 December 1833 and submitted a decidedly luke-warm report. It would be of benefit only to local residents and there was no justification for it on Post Office grounds. The road was bad and hilly and so the contract would be expensive: the lowest tender he had been able to obtain was £7 10s per mile, or £180 *p.a.* The only category of mail to benefit would be that to Merthyr from places to the west of Neath, including Ireland, which produced a total annual income of less than £130. In any case much of the correspondence relating to Merthyr would continue to go to Cardiff, since that was where the agents of the ironmasters were based and where the iron was actually sold. Local correspondence between Neath and Merthyr was worth about £119 10s a year, although this could be expected to increase. Since a kind of promise had previously been extracted from the Post Office, Freeling grudgingly recommended that an experimental service be established between Neath and Merthyr, initially for one year only, and that Rideout's proposal to set up a Penny Post from Neath to Glynneath should be accepted.

The Postmaster General agreed to this on 9 January 1834 and steps were taken to implement the new arrangements. A meeting was called

---

[1] *The Cambrian* 31 August 1833, 14 September 1833; *Glamorgan, Monmouth and Brecon Gazette, and Merthyr Guardian* 7, 14 September 1833

at Merthyr on 15 January to explain them, with Guest in the chair. Among those present was William Williams, who protested that his interest in the proposal was not merely a matter of personal convenience: ' ... he and his neighbours received at present their letters as soon, and somewhat cheaper, than they would under the proposed plan', which probably confirms the existence of a private messenger from Neath to Glynneath. Another speaker at the meeting, a D.W. James, made the point that good communications between Merthyr and Ireland were important because they affected the supply, and consequently the prices in Merthyr, of provisions from that country, which is an interesting insight into how this town of over 22,000 inhabitants, was provisioned.[2] It is also worth noting that the Merthyr ironmasters sold a significant proportion of their product to Ireland: in 1822, for instance, Dowlais received over 400 letters from addresses in that country, representing nearly 22 *per cent* of the total number of incoming letters.[3]

The Neath-Merthyr mail cart and the Penny Post, 1834-1835

The new service was provided by a mail cart and started on 17 February 1834. It left Neath each day at 7.00 a.m. after the arrival of the coach from Swansea and was due in Merthyr at 10.30 a.m. In the opposite direction it left at 2.30 p.m. and was back in Neath at 5.00 p.m. The contractor was a Mr Luce, described as the clerk at Swansea post office.[4] Luce did not last very long. On 19 April Freeling reported that he was unable to continue because he had lost some of his horses and recommended that the contract be transferred to Purchase and Price at £180 *p.a.* Edward Purchase was the landlord of the *Castle Inn* in Merthyr, the scene of the confrontation between the military and the rioters on 3 June 1831.

At the same time as the mail cart started, the Glynneath Penny Post was set up. Its first day of operation, 17 February, must also be the date when the receiving house was opened. A bag for Glynneath was made up at Neath and received there about two hours after the departure from Neath. It contained letters from all parts of the country. The cart picked up the outward letters on its return to Neath the same afternoon. Mail

---

[2] *Glamorgan, Monmouth and Brecon Gazette, and Merthyr Guardian* 18 January 1834
[3] Glam RO DG/A/1/82-92
[4] *The Cambrian* 31 January 1834

going westwards then carried on through Swansea with little delay, but mail to London and the rest of the country, including Merthyr, had to wait overnight at Neath until the up coach arrived at 6.45 a.m. the following morning. It was estimated that the proceeds of the Penny Post would be £20 *p.a.* and the outgoings were put at £6 *p.a.* for the Receiver's salary. This projected income equates to 4,800 letters a year or about thirteen a day. In fact the results turned out to be a lot better than had been anticipated: in the year ended 5 July 1836 the income was £33 1s 6d, or 7,938 letters.

Swansea-Merthyr mail coach, 1835-1851

The results of the first twelve months of the Neath-Merthyr mail also exceeded expectations. The total income for the year was £299 6s 3d, which was £119 6s. 3d above the cost of providing the service, and there was no hesitation in recommending that it be continued. From 6 July 1835 the mail cart was upgraded to a mail coach and extended back to Swansea. It left Swansea at 7.00 a.m. and arrived in Merthyr at 11.30 a.m., slightly later than before. In the opposite direction it left at 2.00 p.m. and was due back in Swansea at 6.30 p.m. In both directions the times were designed to provide a connection at Swansea with the up and down Irish mails.

It soon appeared that earlier misgivings about the state of the road between Neath and Merthyr had been fully justified. After little more than six months, in January 1836, the contractors were talking of giving up the coach: ' ... owing entirely to the objectionable state of the present road ... it is found by the contractors to be too arduous an undertaking'.[5] This resulted in urgent attempts to improve the road so as not to lose the benefits of the mail coach. The Neath turnpike commissioners applied for a government loan, and sought financial support towards the guarantee to the tune of £2,500 in Swansea. Fund raising proceeded and by November the turnpike trust was in a position to go ahead. The section of road that was most in need of improvement was that between Glynneath and Merthyr where the road had to climb out of the Vale of Neath and up to Merthyr across rough and inhospitable country. Ultimately the road was improved, but not before Purchase & Price had surrendered their contract. They gave formal notice of their intention in

---

[5] *The Cambrian* 30 January 1836

November 1836 and it took effect from 5 April 1837. Immediately rumours flew about that the Post Office was about to abandon 'this important commercial connection', as it was described, but the rumours were to be proved groundless. In March 1837 the Post Office advertised that it was ready to receive tenders ' ... for running a pair horse mail coach between Swansea and Merthyr, by way of Neath and Hirwaun'.[6] A new contract was duly awarded to Crouch, who also had the contract for the Abergavenny-Merthyr and Cardiff-Merthyr coaches. The cost was 3d per double mile (i.e. there and back) over a distance of 32 miles, and so the annual value of the contract was £146, considerably less than that had been paid to Purchase & Price. The new contractors probably took over at the beginning of the summer of 1837.

In September 1836 Rideout had been asked to investigate complaints from Glynneath over the level of service that was provided under the existing arrangements. The complainants were probably more interested in a service to Cardiff and London than to Ireland or Merthyr and these needs were not being met. Letters to London had to go the long way round through Neath, and this added to the cost of postage. It also meant that there were only a few hours in the middle of the day to reply to a letter by return – the time, in fact, that it took the coach to get to Merthyr and back. Rideout reported that if Glynneath were to be made a Penny Post under Merthyr as well as under Neath, then letters from London and the north could be routed via Merthyr and they would arrive the previous evening (always assuming that a messenger were employed to carry the letters to Glynneath immediately after the arrival of the London mail at Merthyr). This idea was accepted and the new Penny Post appears to have been implemented. Control of Glynneath probably passed from Neath to Merthyr at the same time. The Penny Post did not last very long, and was withdrawn on 5 August 1837.[7] It is not clear what effect this had on the circulation of letters, although Glynneath seems to have remained under the control of Merthyr.

Perhaps as a reflection of the state of the road, it was not uncommon for the mail coach to be temporarily discontinued during the winter months, or for its frequency to be reduced. Thus in the winter of 1837/38 and again in 1841/42 it was suspended from the end of

---

[6] *The Cambrian* 4 March 1837
[7] Archer, Blakely and Jones 1987,62

October or early November until the following Spring; and in the winter of 1840/41 the service was reduced to thrice-weekly, Monday, Wednesday and Friday only. What happened to letters when the coach was not running is not clear: did they revert to the original route via Cardiff, or was a mail cart substituted?

Further adjustments were made to the timings over the following years which reflected changes to the times of the Milford Haven mail coach through Swansea. In 1840 arrival at Merthyr was at 12.30 p.m. and despatch was at 3.45 p.m. By 1841, however, the times were:

| Swansea | 8.20 a.m. | Merthyr | 2.45 p.m. |
| Merthyr | 12.50 p.m. | Swansea | 7.15 p.m. |

A single coach made one return journey daily, and no Post Office guard was carried; the letters were in the care of the driver.[8]

In 1842 the departure from Swansea was advanced and the times became approximately:

| Swansea | 6.15 a.m. | Merthyr | 1.00 p.m. |
| Merthyr | 10.30 a.m. | Swansea | 5.15 p.m. |

The changes to the times of the coach, combined with improvements to the coach from Abergavenny to Merthyr carrying the London and North mails, which was now due to arrive at 1.00 p.m., meant that there was now a connection at Merthyr between the two. In the opposite direction the Abergavenny coach departed at 11.00 a.m. John Henry Vivian's vision of an Abergavenny to Swansea mail coach was now almost achieved. The Neath coach was not used for mail from London to west Wales, but it did mean that Hirwaun and Glynneath could receive their London and North mails a day earlier.

By 1848 a short-lived private post was in existence to Banwen ironworks, presumably from Glynneath. Morgan Evan, described as 'postman to the Banwen Iron Company', was charged at Neath Petty Sessions with assaulting a farmer, Griffith Griffiths. It turned out that the complainant was the real aggressor, and that he was drunk when the fracas took place. The case was dismissed.[9] Banwen ironworks was one of a number of works set up in the 1840s in an attempt to smelt iron

---

[8] 'Return of mail guards' 1841
[9] *The Cambrian* 17 March 1848

with the local anthracite. It only lasted from 1846 to 1849 and less than 100 tons of iron were ever produced.[10]

Hirwaun receiving house

In 1843 a second receiving house on the road between Neath and Merthyr was established at Hirwaun. In view of the level of industrial activity at Hirwaun and the size of the population it is surprising that an official post had not been set up at an earlier date: a Penny Post from Merthyr might well have been expected. The decision to set up a post office was taken on 7 January 1843 and the first receiver, David Davies, was appointed on 27 May. A handstamp was issued on 10 June and presumably the office was in business soon afterwards. It was set up on the basis that expenses were to be guaranteed by the inhabitants, which suggests that the Post Office did not expect very much business from the place. They were probably right, for even in 1845, when about 100 letters a week were passing through the office on average, this was only about two-thirds of the number that had been handled at Glynneath ten years earlier. 100 letters a week was judged to be enough to justify making the office a charge against the Revenue, which was done from 6 April 1845. The only outgoings were £5 *p.a.* for the receiver's salary. There was no house-to-house delivery.

Hirwaun was made dependent on Merthyr. This had the unfortunate effect that, going east, the mail coach passed through Hirwaun without putting off any mail even though it stopped to collect the outgoing letters. Mail from Swansea and the west was carried on to Merthyr and had to wait there until the following afternoon when it was sent back to Hirwaun on the west-bound coach. Similarly mail for Swansea had first to go to Merthyr and then return through Hirwaun. In October 1845 and again in October 1846 requests were made to have direct bags to Hirwaun made up at Swansea and Neath. This would have enabled mail from the west to be put off at Hirwaun without having to pass through Merthyr. The request was unsuccessful on both occasions: the Post Office claimed that only about sixteen or eighteen items a week would benefit, and it would create additional administrative work at Swansea and Neath where the postmasters would have to open and maintain accounts with Hirwaun. In May 1847 a further request was made, and

---

[10] Hughes 1990, 25-31

again the Post Office responded that there was not enough traffic to justify the change. However, on this occasion, perhaps by way of a sop, it was agreed to establish a free delivery in the village; the surveyor was directed to propose a boundary and report on the cost. This unsatisfactory situation was still unchanged in 1850 when it was brought to the attention of a wider audience through a letter to *The Times*.[11] It remained unchanged until the abandonment of the mail coach in the following year led to a re-casting of postal arrangements along this route.

The service to Aberdare, with which Hirwaun had closer links than with either Neath or Merthyr, was equally unsatisfactory. The actual distance between the two places is only about three miles, but until 1851 letters had to make a round trip of 54 miles, via Merthyr, Pontypridd and Aberdare, taking two days over the journey.

In 1845 the pattern of mail circulation in south Wales was extensively revised following the opening of the Great Western Railway to Gloucester. Mail was sent as far as Gloucester by rail and then on by coach. The Neath coach was re-timed to take account of the new timings now introduced for the London mails at Swansea and Merthyr. The new pattern required two coaches, passing en route, instead of one making a daily return journey. The approximate times, which were moved a little earlier over the following years, were:

Swansea	10.00 a.m.	Merthyr	10.15 a.m.
Merthyr	2.45 p.m.	Swansea	3.00 p.m.

After the mail coach

The end of the Swansea-Merthyr mail coach came in 1851. Following the opening of the South Wales Railway the Post Office concluded that it could maintain a satisfactory service for less cost if it used the railway from Swansea to Cardiff and then sent the mail on by road through Pontypridd and Aberdare. Both parts of the route were already covered by existing contracts and so there would be no additional cost. Notice to quit was given to the contractors in April 1851 and an advertisement appeared inviting tenders from parties who were willing to operate a mail cart from Aberdare to Glynneath.[12] The coach ran for

---

[11] *The Times* 16 September 1850; *Cardiff & Merthyr Guardian* 21 September 1850
[12] *The Cambrian* 16 May 1851

the last time on 5 July, after which ' ... its respectable Jehu retires on his laurels to a tranquil spot near the Mumbles'.[13]

The service from Aberdare to Glynneath commenced on 6 July 1851, but it would appear that there had been no response to the earlier advertisement for a contractor to operate a mail cart (or at least no acceptable responses), since in March 1852 the service was referred to as a 'ride' or horse post. It operated in connection with the mail cart from Pontypridd to Aberdare which had started a few months earlier. At the same time, and as a consequence of these changed arrangements, control of Hirwaun and Glynneath passed to Pontypridd (which had controlled Aberdare since February 1851). Hirwaun subsequently came under the control of Aberdare in 1855 when the latter was made a post town.

The new arrangements suited Hirwaun well enough. They now received their letters at 11.45 a.m. and were able to send a reply on the same day, if they wished, since the outward mail went at 2.30 p.m. However, they did not go down so well at Swansea, where they were said to be ' ... a source of much discontent, delay, and inconvenience'.[14] Glynneath was even less well pleased. Letters from Neath, just ten miles distant, now had to go 70 miles round by Cardiff, Pontypridd and Aberdare and took two days over it. But in August 1851 the Post Office concluded that there was no way of improving the service except at a greater expense than the level of business would justify.

However, the complaints continued[15] and the Post Office accepted that the existing arrangements were not acceptable. In March 1852 Maberly – obviously without any great enthusiasm – passed on the recommendations of the surveyor: 'Under the circumstances stated in this report from the Surveyor I see no alternative but to discontinue the Aberdare and Glynneath ride and to serve the district in the manner proposed.' Glynneath was to be served by the Vale of Neath Railway which had recently opened from Neath to Aberdare. The railway company proposed a charge of £10 *p.a.* for carrying a bag from Neath to Glynneath every day. The sub-postmaster at Glynneath was to have an additional £2 *p.a.* for delivery and £5 for carrying the bag to and

[13] *The Cambrian* 11 July 1851
[14] *The Cambrian* 10 October 1851
[15] *The Cambrian* 23 January 1852

from the station. The disparity between these sums must reflect how few letters there were to be delivered each day and what a compact area the village of Glynneath then occupied.

Hirwaun was to be served by extending the existing foot post between Aberdare and Trecynon. Since the local delivery would now be carried out by this letter-carrier, the salary of the sub-postmaster at Hirwaun was to be reduced by £4 *p.a.* These arrangements were approved on 12 March 1852 and probably came into force on 23 March 1852, which is the date on which control of Glynneath passed from Pontypridd to Neath. These arrangements, at both Hirwaun and Glynneath, were still in force when Trollope carried out his review of rural posts in 1852-3. By 1859 the lower part of the Neath valley also was being served by a foot post to Aberdulais, Ynysygerwyn, Abergarwed, Resolven and Melincwrt.

	Appointed	In office	Resigned
Receivers at Glynneath			
??	1834		
John Thomas		1840, 1857	
Receivers at Hirwaun			
David Davies	1843	1848	
Edward Morgan Edwards		1850	
Susanna Williams		1852	
John Williams (grocer)		1856-8	

Chapter 9

Pontypridd and the Rhondda valleys

Pontypridd's position at the confluence of the rivers Rhondda and Taff made it a focal point for the Rhondda, Taff and Cynon valleys and consequently an important commercial and administrative centre within the coalfield as well as an industrial town in its own right.

The development of Pontypridd, or Newbridge as it was originally known, goes back no further than the middle of the eighteenth century. In 1756 William Edwards erected his famous bridge across the Taff at a point which was already favoured as a crossing because of the comparative shallowness of the water. Until 1771 the road from Cardiff to Merthyr ran through Caerphilly and Gelligaer, but in that year a new turnpike road was started from Tongwynlais to Merthyr. It kept close to the eastern bank of the Taff and passed through Pentrebach on the opposite side of the river to what became Newbridge. If there was any sort of settlement at all at this date around Edwards' bridge it was too insignificant to be worth a detour.

The development of Newbridge began in the 1790s with the construction of the Glamorganshire Canal. Newbridge, conveniently located at the halfway point on the canal between Cardiff and Merthyr, was well placed to develop as both a service centre for the mid-valleys area and as an industrial centre. The canal provided good transport links; flat land, suitable for industrial development, was available where the Taff valley opens out as the river emerges from the hills; and coal was available in the immediate vicinity. All these factors encouraged industrial and commercial development. William Crawshay I opened a forge and nail works at Ynysangharad in 1800. In 1818 the site was acquired by Samuel Brown and his cousin, Samuel Lenox. They erected the Newbridge Chain Cable and Anchor Works where they manufactured their patent stud-link wrought-iron chains. The company grew rapidly in size and reputation and became sole supplier of anchor chains to the Royal Navy. Brown Lenox became synonymous with Pontypridd and contributed greatly to the town's successful growth. The works closed in 2000, latterly manufacturing rock-crushing machinery rather than chains.

A second focus of industrial growth was at Treforest, about 1½ miles to the south-east, where a small tinplate mill had been erected in the 18th century. It was acquired by William Crawshay I in 1794. His son, William Crawshay II, expanded and modernised it and eventually placed it under the management of *his* son, Francis Crawshay. By 1836 it was claimed to be the largest tinplate works in Britain.[1] These two undertakings in particular resulted in Newbridge becoming an important manufacturing town. The combined population of Newbridge and Treforest had reached about 2,000 by 1835.

Slightly later, from about 1840, the Rhondda coal industry entered a phase of rapid development. The first attempt at exploiting the coal reserves on a commercial basis was in 1809 when Walter Coffin opened levels at Dinas, about five miles west of Newbridge. Coffin's venture was successful: in 1830 he shipped over 46,000 tons of coal and in 1840 this figure had risen to nearly 51,000 tons, but until the 1840s his was the only significant operation in the valley. The opening of the Taff Vale Railway to Dinas in 1841 encouraged further development. In the following decade collieries were opened all along the valley between Newbridge and Dinas, including the first colliery in the Rhondda Fach at Ynyshir. The coal produced by these collieries was bituminous and no attempt had yet been made to exploit the steam coal which was believed to lie under the upper reaches of the two valleys. In 1854 these reserves were proved at Treherbert, nearly at the head of the Rhondda Fawr, the railway was extended and commercial production started in 1855. This resulted in the rapid opening of pits along the whole length of the two Rhondda valleys and, as a consequence, in the construction of densely packed terraced housing on the valley floors which eventually stretched from Pontypridd to the heads of both valleys with virtually no interruption. This growth of population in the Rhondda valleys contributed greatly to Pontypridd's success as a regional centre. In the same way as Merthyr is the classic iron smelting town, and can be seen

---

[1] 'Rhondda Cynon Taff Libraries Heritage Trail'). Available from world wide web: http://webapps.rhondda-cynon-taf.gov.uk/heritagetrail/taff/treforest/Treforest.htm (accessed on 4 December 2006)

as the exemplar of the first, iron-based wave of the industrialisation which started in the 1760s, so Rhondda is the classic coal mining settlement, typical of the communities which developed during the second, coal-based wave of industrialisation which began in the 1840s.

~~~~~~

Private receiving house at Pentrebach

By 1742 there was already some sort of regular but unofficial post between Merthyr and Cardiff but the messenger most likely followed the original road through Caerphilly and Gelligaer. Once the turnpike from Tongwynlais along the present-day route was constructed in 1771, he presumably took to the new road. By the time Samuel Woodcock compiled his notes on the district in about 1790 there was a regular private delivery from Cardiff to Newbridge which was perhaps carried out by the same messenger who maintained the three-day post between Cardiff and Merthyr (see chapter 4).

Woodcock also recorded a private delivery from Cardiff to 'Lanwona' which is probably Llanwonno, an upland parish bordered by the rivers Rhondda to the south and Taff to the east. What is now the town centre of Pontypridd lies in the south-eastern corner of the historic parish. In the 18th century the Griffiths family of Gellifendigaid, about two miles north of the present town, were the principal landowners. A member of the family, Dr Richard Griffiths (1756-1826) is generally regarded as one of the pioneers of the Rhondda coal industry. He did not work the coal himself but built a tramroad from Gyfeillon to the Glamorganshire Canal at Pontypridd in 1809 which made it possible for other entrepreneurs, such as Walter Coffin, to open up their sale coal levels. Woodcock's Lanwona post probably represents an extension to Gellifendigaid of the Cardiff-Newbridge private post noted above.

In 1804 an official horse post was started from Cardiff to Merthyr and this was upgraded to a mail coach in 1821. There was still no official post office in Newbridge but by 1830 a private receiving house existed, where exactly is not known. The mail coach did not enter Newbridge, since it was impossible for any vehicle to cross Edwards' steeply-arched bridge, but it stopped at the *Duke of Bridgwater Arms* in Pentrebach to change horses. The landlord of this inn, Thomas Newman, was one of the shareholders in the company which horsed the coach, the others being Philip Woolcot, of the *White Lion Inn*, Cardiff and Mr Trehern of the *Castle Inn*, Merthyr. Letters for Newbridge in a separate bag were handed over to the private receiver during the change of horses. The *Duke of Bridgwater* is no longer in existence, but its site was close to Glyntaff church alongside the main road.

A collection of 23 letters of this period from Brown, Lenox, the Newbridge chain manufacturers, to the Dowlais Iron Company is preserved among the Dowlais letters.[2] With one exception none of them bear any kind of postal marking. They all appear to have been sent under cover since there is no address on the verso, and they may have been carried by private messenger. The one item that was definitely carried by post has a Cardiff mark dated 25 DE 1822 – but who is to say whether it entered the post at Newbridge or at Cardiff?

In 1830 the keeper of this private receiving house and his wife were disgraced and the opportunity was taken to set up an official sub-office. The problem was first brought to light in a letter from one Adam Rowland to Sir Francis Freeling. He complained that in certain cases the postage on letters he had received from Swansea had been altered from 7d to 8d.

Freeling passed the matter on to the local Surveyor, Charles Rideout, to investigate. Rideout made enquiries at every stage along the route between Swansea and Newbridge, culminating in what was obviously an intensive interrogation of the receiver's wife. It is interesting that initially he did not rule out the possibility that Rowland had altered the rates himself in pursuit of some private dispute with the receiver. He reported back to Freeling a couple of weeks later:

> Carmarthen 15 Nov 1830
>
> Sir,
>
> I beg to return the enclosed from Mr Adam Rowland of Newbridge near Cardiff complaining that the rate of postage on his letter had been altered.
>
> It appeared to me evident that the tax had been improperly altered & by a person not connected with official duties. I therefore proceeded to Swansea where

---

[2] Glam RO DG/A/1/83

the letter was put in, & from the Deputy learnt that the 7d on the enclosed letter was taxed by his assistant, that the charge was not altered there & that all letters addressed to Newbridge were charged the rate to Cardiff only namely 7d. The Postmr of Cardiff & his assistant denied all knowledge of the alteration. I then proceeded to Newbridge & had an interview with Mr Adam Rowland who not only asserted that the cover (No 1) was altered previous to its delivery, but in answer to my question produced two other covers similarly altered (No 2 & 3) one of which bears the Swansea stamp of Feby 21, 1827. Mr Thomas of the chain cable manufactory Newbridge on whom I called relative to an official proposition, which will be contained in a separate report, informed me that he had some years back discovered a like transaction against the same parties, but for some reason or other did not bring it forward.

Newbridge is situated out of the Mail Coach line from Cardiff to Merthyr; the letters are carried in a private Bag by the Mail Coach & left at a Mr Kiffs at Eglwysilan, which is on the road near to Newbridge. I questioned the wife of this private receiver & who has the care of the letters, relative to these letters. She admitted that the cover No 1 was altered previous to its delivery but denied its having been altered at her house & hesitated when I asked her if it was received in its present state from Cardiff altho' I put the question to her at three different times. There appears to have been some dispute between this woman & Adam Rowland with respect to the payment of postage. Still her statement will acquit the latter of having made the alteration in the tax with the object of bringing forward a complaint.

I have little or no hesitation in stating that the alteration of the taxes on all the enclosed covers took place at Newbridge & that the private Receiver has been in the habit of committing these frauds for some years. The proposition contained in the accompanying report will place Newbridge, if sanctioned, on a very different footing & the Postmr of Cardiff can appoint a more respected person to take charge of the office in that neighbourhood.

Although Rideout does not go into detail, it is clear that what the receiver had been doing was to alter the manuscript markings which had been applied to letters at the originating offices to indicate the postage to be collected on delivery so that they showed a higher sum. He was then collecting this amount but only remitting to the postmaster at Cardiff the original amount and keeping the difference himself.

The 'accompanying report' mentioned in Rideout's letter was also written at Carmarthen and on the same day. In it he recommended the establishment of an official sub-office which would not only generate revenue for the Post Office but would also provide a cheaper service for the public. It would also ensure proper control over the receiver. The sub-office would be in Pentrebach, on the route of the mail coach and in the parish of Eglwysilan, but not in the village of Eglwysilan itself.

Carmarthen 15 Novr 1830

Sir,

When at Newbridge relative to the complaint made by Mr Adam Rowland it occurred to me that the situation of the place & the Revenue would warrant some official Post & that the existing arrangements could not be maintained without a post of some kind.

Newbridge is situated about ¾ of a mile from Eglwysilan, which is on the Mail Coach line from Cardiff to Merthyr Tidvil & 11 miles & a quarter from Cardiff & about 13 from Merthyr Tidvil. Mr Thomas of the Chain-cable Works has a private Bag from Cardiff whose postage is about £2 per month. The Letters consigned in the private receiving House Bag is about £4.4. a month & the short letters about £1.10 monthly, making a revenue of nearly £100 per annum. It has been the custom at Cardiff to charge all letters put in there to be conveyed in the private Bags to the Receivers or to the different Works, the four Penny rate of Postage & accounting for the same to the office under the head of "Short Letters in private Bags". It occurred to me that as Eglwysilan is nearer to Cardiff than to Merthyr that it would be considered in the delivery from the former Town, consequently the charge of 4d on the short letters could not perhaps be maintained, which short letters average £18 per annum. In observing the letters to Eglwysilan I found they were principally from London, Bristol, Newport and Swansea. In adding the distance from Cardiff to Eglwysilan, I found that the rate from Newport would be 2d more, & from London, Bristol & Swansea 1d more than to Cardiff, in consequence of this it suggested itself to me that a sub office would be in every respect advantageous to the Revenue and to the Public inasmuch as it would reduce the extra charge now made.

At Eglwysilan there are but 4 houses which I submit will be all that can be encluded [*sic*] in the free delivery. Newbridge is at least three quarters of a mile from it & unconnected with it. Pentrebach is about one mile the Cardiff side of Eglwysilan which will have the benefits of the arrangements I am about to submit. The Inhabitants of these Villages can obtain their Letters from Eglwysilan for the legal postage only, but if they be delivered, I trust, the Sub Deputy will be allowed to charge 1d for his trouble, which sum is half what is now demanded. I shall now beg to submit a salary of £10 per annum, for a Sub Deputy at Eglwysilan, that amount, to include the expense of a free delivery to the 4 houses above named.

If I have been sufficiently clear in this statement it will be seen that the expense of making Eglwysilan a Sub Office will but little exceed half the amount of the short letters – & we may fairly look for an encreased [*sic*] revenue on the whole correspondence.

If this proposition be adopted I shall beg to propose, at the suggestion of Mr Thomas, a Bag & short letter account between Merthyr Tidvil & Eglwysilan.

I submit this with all deference ...

Freeling was well satisfied with Rideout's investigations and recommendations. On 17 November he forwarded them to the Postmaster General, no more than a formality, who duly gave his approval and steps were taken to create the Eglwysilan sub-office at Pentrebach on the terms proposed by Rideout.

Eglwysilan official receiving house, 1830

The first official receiver or sub-deputy at Eglwysilan was Thomas Morris. According to Wilkins, he was a grocer who originated from Merthyr.[3] In Pigot's *Directory* for 1835 he appears as a carrier who operated a service to Cardiff from his house twice a week. By 1836 he had resigned his position and moved to Caerphilly. At the time of his appointment in 1830 Rideout reckoned that the office would handle business to the value of about £100 *p.a.* At an average of 6d per letter, this would represent about 4,000 letters a year. In 1843 there were said to be about 400 letters a week, or more than 20,000 a year at 1d each.

When the office was first set up the sub-deputy was given a salary of £10 *p.a.* This included payment for making a free delivery within the town which in practice was confined to a very restricted area. Complaints were made in 1839 that it extended for no more than 300 yards from the office. There is no mention of a letter-carrier being appointed so presumably the sub-deputy was expected either to carry out this task himself or pay someone to do it. This rate of remuneration seems to have been normal for a sub-office that had to carry out a free delivery. Where this was not required the sub-deputy received as little as £3 or £4 *p.a.* The first letter-carrier is said to have been one Roberts who wore spectacles 'straight from nose to forehead'.[4] By 1846 the letter-carrier had been made an official appointment, for on 14 April he made an unsuccessful request for an increase in wages. His request is understandable in view of the increased number of letters but 10s a week seems to have been the normal weekly rate for a letter. There were two daily deliveries in Pontypridd at this time – the London and Western mails at 10.00 a.m. and the North mail at 2.30 p.m.

In 1835 letters to the south and west via Cardiff were despatched at 8.15 a.m. whereas letters to Merthyr and north Wales (which probably means

---

[3] Wilkins 1888, 144
[4] Wilkins 1888, 144

the North Mail in general) were despatched via Merthyr at 4.15 p.m.[5] Incoming letters were probably received in the same way. This makes it clear that letters were routed through either Merthyr or Cardiff as appropriate, reflecting Rideout's recommendation that an account be set up between Merthyr and Eglwysilan. Even though Eglwysilan was a sub-office under Cardiff, the existence of this account ensured that letters did not all have to be despatched through Cardiff regardless of their destination.

In 1846 the Taff Vale Railway started to carry mail between Cardiff and Merthyr, and naturally Pontypridd benefited from this development. The London and Western mails came up the valley by train, while the North mail continued to come down from Merthyr. The same train that brought the North mail in at 2.21 p.m. also took the outward London mail down to Cardiff, and *vice versa* in the morning at 9.35 a.m.

Wilkins gives the impression that Thomas Morris was succeeded directly by Charles Bassett as sub-deputy at Eglwysilan. However, in Post Office Archives and the contemporary press there are references to a number of short-lived appointments between these two. D. Davies, described as 'postmaster, of Newbridge' was married in May 1836[6] and on 29 October 1839 John A. Smith was appointed to 'Egliosilian'. On 27 April 1841 the resignation of the receiver of Eglwysilan was minuted: no name is given, but presumably it was Smith. He was followed by D.N. Thomas who was appointed on 17 September 1841. He appears to have resigned almost immediately because on 3 November 1841 there was another appointment, of John Evans to 'Eglwyswern', presumably Eglwysilan, since no other place with a name resembling this had a postal service at the time.

Post office moved into Newbridge, 1841

D.N. Thomas probably had a business in Newbridge itself, for on 1 October 1841 it was minuted that the office was to be moved from Eglwysilan to Newbridge. The decision to move the office was followed by the issue of the first recorded postmark, an undated circular handstamp reading 'Pontypridd' which was proofed on 20 November

---

[5] Pigot 1835
[6] *The Cambrian* 11 June 1836

1841.[7] Moving the post office into Newbridge from Pentrebach was long overdue. The fact that the post office in what was described as a 'rapidly increasing manufacturing town' could be 'kept at an inn, nearly a mile from the town' was cited in 1839 as one of the evils arising from the 'want of management' in the Post Office'.[8]

By the time the new office opened Thomas had resigned and John Evans had taken over. He in turn resigned after about a year and was replaced by Charles Bassett on 11 October 1842[9] (not 1843, as Wilkins has it). Bassett made a name for himself in the wider community during his time as postmaster of Pontypridd. Like Charles Wilkins at Merthyr, he is one of the few figures in the Post Office in Victorian south Wales who takes on a personality and is not simply a name in the historical record. Bassett was born at Siggiston in the Vale of Glamorgan in 1818. His father was the Vicar of Colwinstone. He moved to Newbridge as a young man in about 1840 and set up as a druggist and chemist, the first in the town. He was appointed sub-deputy in 1842 and very soon established himself as a leading figure in the commercial and public life of the town. In 1844 he gave evidence to the Royal Commission appointed to investigate the causes of the Rebecca Riots. He was later responsible, with others, for establishing the first Wesleyan chapel in the town; he was a member of the committee that built the Victoria bridge to replace Edwards' original bridge of 1756 and was a founder member of the Pontypridd Market Company. He resigned as postmaster in 1874 and became manager of the local branch of the London & Provincial Bank. He died in Pontypridd at his home, *Brynffynon*, on 1 April 1887.[10]

The appointment of such an able young man as sub-deputy was a fortunate one for the coming years saw a great expansion in the amount of business transacted at Pontypridd and in the responsibilities of the sub-deputy. One of the first of these changes was the formal alteration of the name of the town from Newbridge to Pontypridd. This was done in

---

[7] Broomfield 1996, 75
[8] *Bristol Mercury* 13 April 1839
[9] POST 58/70
[10] This account of Bassett's career is based on Charles 1920. For a wider account of his family, see Rees 2008. See too his obituary in *Western Mail* 2 April 1887. For a portrait of Bassett, see 'Gathering the Jewels', available from: http://www.gtj.org.uk/en/small/item/GTJ69047/ [accessed 5 January 2010]

*The grave of Charles Bassett (postmaster of Pontypridd, 1842-74)
in Glyntaff cemetery, Pontypridd*

1845 at the instigation of Bassett in order to avoid confusion with places
such as Newbridge-on-Wye or Newbridge near Newport:

> The name of Newbridge in Glamorganshire is to be
> altered to Pont-y-pridd. This is the suggestion of the
> General Post-office. The letter stamp is now altered to the
> above. There were numerous mistakes in sending letters
> to Newbridge near Pontypool. It is in contemplation to
> alter the name of Newport, in Monmouthshire, to

Uskport. There are no less than six towns in Great Britain bearing the name Newport.[11]

Despite what was written in this report, the 'letter stamp' had read 'Pontypridd' since 1841 when the office was moved into Newbridge from Eglwysilan.

In December 1846 a request was made by the inhabitants for Money Order Office status. This was granted, although there was a delay in implementing the decision and the new status did not take effect until July 1848. Maberly noted that correspondence averaged 800 items a week (which was a growth of 100 *per cent* in less than four years since 1843) and anticipated that M.O.O. status would lead to a lot of additional business: 'I fear Your Ldsp must make some considerable addition to the Sub-Deputy's salary at present only £12 a year', he wrote to the Postmaster General. In view of the great increase in the level of business that had already taken place in a very short time and the further anticipated growth (which fully materialised), this seems to have been a reasonable proposal. The Postmaster General was convinced and more than doubled Bassett's salary to £25 *p.a.*

Post town status, 1850

In 1849 when Bassett was re-designated sub-postmaster. The following year, on 6 October 1850, Pontypridd was made a post town as part of the revised arrangements that were required following the reversion of the Merthyr mail from the railway to a mail cart as described in chapter 5.

> PONTYPRIDD. On and after the 6th of October next, the
> Post-Office at this fast-increasing town will be a
> Principal Office, having direct communication with
> London, Bristol, and other towns, and not, as heretofore,
> a Sub-Office under Cardiff. It will consequently be
> necessary that, in future, all letters for the town and
> vicinity be addressed "Pontypridd, Glamorganshire", not
> near Cardiff, to avoid 24 hours' delay. The appellation
> "Newbridge" should also be entirely dropped, or constant
> delay and irregularity in the transmission of letters will
> occur, there being in almost every county in England and

---

[11] *The Cambrian* 9 August 1845

Wales one or two places called by that name. It is hoped that the Taff Vale Railway company will also change the name of the station. Since the appointment of the present Postmaster, in October, 1842, the letters have increased four-fold. The Office was made a Money Order Office in July, 1848.[12]

Following its establishment as a post town, Pontypridd took over control of a number of existing sub-offices and several new deliveries were established in the following years.

Treforest. A receiving house had already been established here in 1847, served by a foot post from Pontypridd. A handstamp was issued on 16 February. Control passed from Cardiff to Pontypridd in 1850 as soon as the latter became a post town.

Nantgarw. A sub-office under Llandaff had been set up in 1846. A handstamp was issued on 30 June and John Bolton was appointed receiver on 9 July. A letter-carrier named Protheroe was appointed in October 1849. Control passed from Cardiff to Pontypridd in April 1852 when the Treforest foot post was extended south to Nantgarw. The increased cost was put at £32 11s 8d *p.a.*, but it was expected that an additional 213 letters would be carried each week resulting in a total income of £46 2s *p.a.* In 1853 the route was given as Treforest, Rhydvilan, Upper Boat, Nantgarw, Dyffrynfrwdd. It was a daily service.

Dinas. Pontypridd became the post town for the Rhondda on 6 October 1850. The first receiving house, at Dinas, was opened at much the same time. A full account of the postal service in the Rhondda valleys is given below.

Nelson. A receiving house was opened late in 1850. A handstamp was issued on 8 November and on 9 December the Surveyor was instructed to select a suitable person to fill the vacancy. A messenger was appointed, three days a week, who also delivered to Navigation (present-day Abercynon) and Quaker's Yard. The cost of the messenger was put at £15 12s a year (i.e. 6s per week) and the receiver at Nelson received £2 a year. Trollope's review of the Pontypridd rural posts in December 1853 gave the route as Penmaen, Pont Shonnorton, Navigation, Nelson.

---

[12] *The Cambrian* 27 September 1850

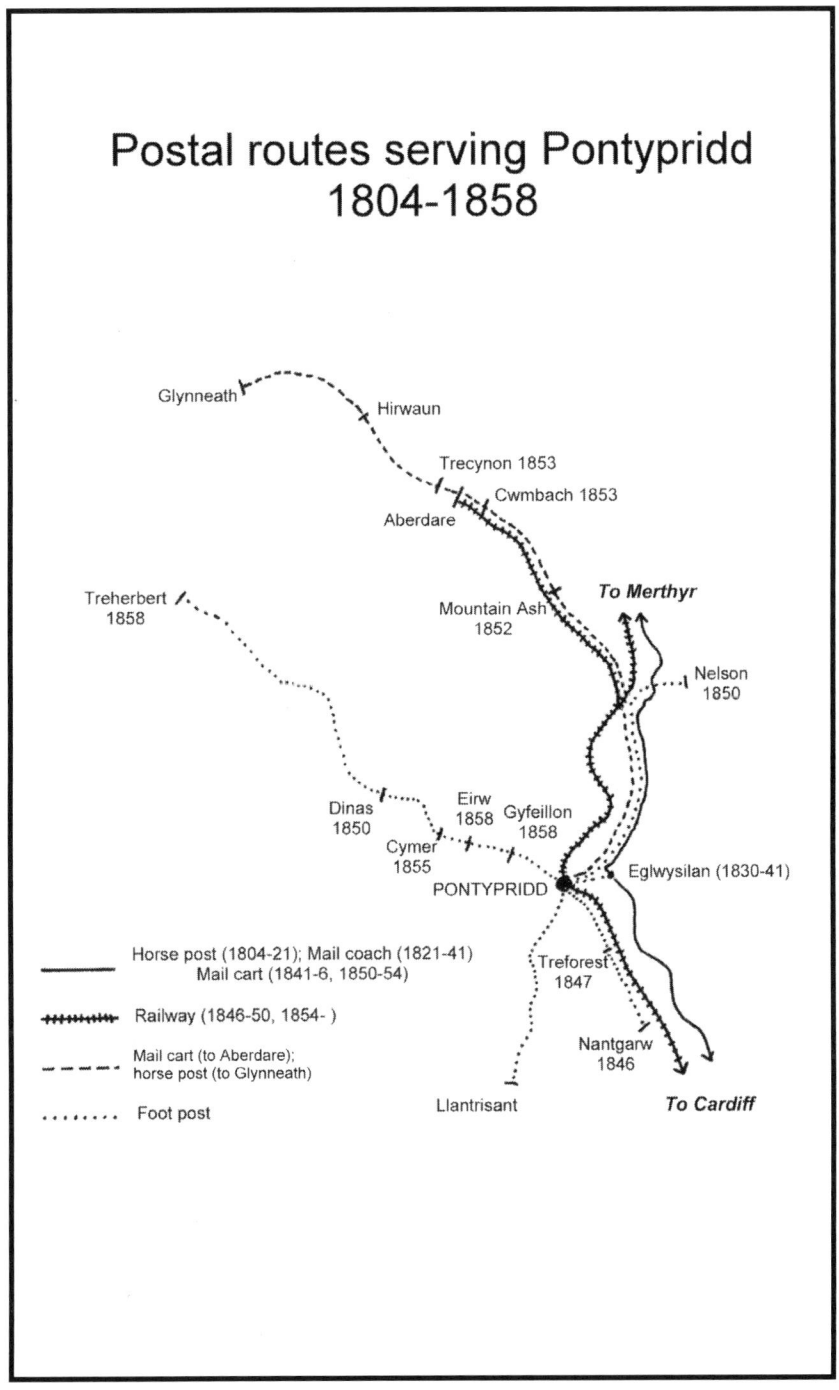

Postal routes serving Pontypridd
1804-1858

Glynneath

Hirwaun

Trecynon 1853

Cwmbach 1853

Aberdare

To Merthyr

Treherbert
1858

Mountain Ash
1852

Nelson
1850

Eirw
1858   Gyfeillon
1858

Dinas
1850

Cymer
1855

PONTYPRIDD

Eglwysilan (1830-41)

Horse post (1804-21); Mail coach (1821-41)
Mail cart (1841-6, 1850-54)

Railway (1846-50, 1854-)

Mail cart (to Aberdare);
horse post (to Glynneath)

Foot post

Treforest
1847

Nantgarw
1846

Llantrisant

To Cardiff

In July 1854, on the proposal of Rideout, the frequency was increased to six days a week; the receiver's salary was increased to £3 *p.a.* and the messenger's wage to 11s 6d a week.

Llantrisant. Control passed from Cardiff to Pontypridd on 6 October 1850. A full account is given below.

Aberdare. Control of Aberdare and its sub-office at Aberaman were transferred to Pontypridd from Merthyr on 10 February 1851, when a mail cart from Pontypridd replaced the foot messenger from Merthyr. Sub-offices were opened at Mountain Ash in December 1852, at Cwmbach in April 1853 and at Trecynon in July 1853. Aberdare became a post town in 1855 and took over responsibility for these three offices.

Glynneath, Hirwaun. Control passed from Merthyr to Pontypridd on 5 July 1851 when the Aberdare mail cart was extended to Hirwaun and Glynneath following the withdrawal of the Swansea-Merthyr mail coach. Full details are given in chapter 8. Glynneath was subsequently transferred to Neath on 23 March 1852, while Hirwaun came under the control of Aberdare when the latter was made a post town in 1855.

In view of the extra work involved in his extended duties Bassett was granted a further salary increase in September 1851 from £25 to £40 *p.a.* He was also granted an allowance of £15 *p.a.* to help with despatches.

Llantrisant

Llantrisant is an ancient borough which is strategically located on a hilltop on the border between the Vale of Glamorgan and the uplands. The town itself lay just outside the coalfield and remained totally unaffected by the industrial developments of the 19th century, although the population of the wider parish of Llantrisant increased because it contained part of the lower Rhondda valley, including the important early settlements of Dinas and Cymer.

Woodcock records a private delivery from Cardiff to Llantrisant in about 1790. There was also a delivery to Castella, a gentry house 2½ miles to the north of the town, and probably the same messenger covered both deliveries. By 1825 Llantrisant had a Penny Post under Cardiff, served three days a week by a messenger from Llandaff, where he met the coach from Cardiff to Merthyr. These arrangements had existed since at least 1813.[13] In 1830 the Revd J.R. Williams submitted a request on behalf of the inhabitants to have a daily post, but Rideout reported that there was nothing like enough business to justify it; in fact any sort of service to Llantrisant was only possible because almost the entire cost of the Llandaff-Llantrisant post was covered by the income from Llandaff. In 1832 it was proposed that Llantrisant should be served from Porth-y-glo, on the post road from Cardiff to Merthyr and near to Newbridge, rather than from Llandaff. Nothing seems to have come of this, for in 1838 there was still a foot post from Llandaff to Llantrisant[14] which still operated three days a week. However, in April 1850 the post from Llandaff was replaced by a post from Pontypridd, for the Surveyor was instructed to appoint a messenger for the walk between Pontypridd and Llantrisant. In October 1850, when Pontypridd was made a post town, administrative responsibility for Llantrisant was also transferred from Cardiff. In December 1853 there was a six-day post from Pontypridd at a cost of 11s a week. The sub-postmaster received a salary of £3 *p.a.*

Rhondda

Walter Coffin opened his first levels at Dinas in 1809. Commercial coal working extended as far as Trealaw in 1839 and Tonypandy in 1845, but the character of the upper parts of the two Rhondda valleys (Rhondda Fawr and Rhondda Fach) remained entirely rural until the 1850s. The impression is sometimes given that in pre-industrial times Rhondda was virtually uninhabited. This was far from the case. In 1840 there were at least 140 farms in the two valleys with a population of around 900. However, the farming was poor, mainly sheep and cattle rearing. In the lower part of the valley oats, barley and corn were grown, but for immediate consumption by the local population and their animals and not as a sale commodity. The existence of coal was obviously known and small amounts were taken for local consumption. The hills were thickly wooded and this led to charcoal burning in many parts of the valleys and the remains of platforms where this was carried out have been identified at a number of sites, especially around Dinas, Gelli, Trealaw and Blaenrhondda. Among the places where the charcoal was used was an iron furnace at Pontygwaith in the Rhondda Fach.

---

[13] Archer, Blakely and Jones 1987, 63 record 1813 as the earliest known example of a Llantrisant Penny Post marking
[14] 'Returns relating to mail carts, etc.' 1837/38

Private posts in the Rhondda

Virtually nothing is known of any postal arrangements that may have existed in the valleys before the first official post was established in 1850. The farmers and the parish vestry of Ystradyfodwg can have had little cause for corresponding with the outside world and Woodcock did not record a private post to any point in the Rhondda. When commercial mining came to the lower part of the valley after 1809 the owners and managers must have established some means of maintaining regular contact with their sales agents, their lawyers and their markets, but no record of it survives.

It has been suggested that at one time Merthyr was the post town for the Rhondda valleys. This is based on an entry in the Ystradyfodwg parish vestry book dated 5 August 1826 and which reads: 'To carrying a letter from Merthyr – 2/6d'.[15] However, by his date the private receiving house at Eglwysilan was in existence and one would have expected mail for the Rhondda valleys to have been left there. It is perhaps more likely that there was something special about this letter which made it necessary to employ a messenger for the specific purpose of bringing it from Merthyr: indeed, it may never have entered the General Post at all. The large sum involved and the fact that it was recorded separately suggest that this was not an ordinary item and no conclusions should be drawn from the entry in the vestry book.

The first private post in the Rhondda valley of which there is any trace was operated by one 'Shôn y Gwehydd' (John the weaver). His name was preserved in an oral tradition that was current later in the nineteenth century as having been the first to collect letters systematically in the lower Rhondda.[16] Certainly some sort of private post, perhaps that carried on by Shôn, was in existence in 1848 when the possibility of establishing an official post was considered for the first time. No further action was taken once it was learned that the average number of letters to 'Cwmmer' and Dinas per week was only 56.

Official Rhondda post

Two years later, on 2 September 1850, it was agreed that an official messenger should be established, four days a week, from Pontypridd to

---

[15] Lewis 1959, 113
[16] *ibid*

'Hopkinstown, Glanrhondda, Gyfeillon, Waun-y-euron, Cummer and Dinas', and that a receiving house should be set up at Dinas. The messenger would receive 8s 6d a week and the receiver £3 a year. On 19 September the Surveyor was instructed to appoint the messenger and a handstamp was issued in October. A receiver had still not been appointed by the Treasury by December and so the Surveyor was instructed to select someone suitable. Presumably he did so although there is no record of the name.

In 1851 the Dinas foot post was upgraded from four days a week to six days with the status of a guaranteed post. 'I submit that the usual course adopted in regards to the establishment of Guaranteed Posts may be followed in this case' ran a report of 24 July. The improvement took place and in 1853 the Dinas messenger was recorded as receiving 12s a week.

By 1855 the foot post had been extended beyond Dinas. In Trollope's review of the rural posts in south Wales the route from 'Pont-y-Pridd' is given as 'Hopkinstown, Tymawr, Gyfillon, Gwaunyriew (?), Cwmmer, Troedyrhiw, Ynishir, Hafod Colliery, Dinas, Ystradyfodwg'. The names are slightly out of sequence, but are all highly evocative and full of associations for anyone acquainted with the history of coal mining in the Rhondda. The inclusion of Ynishir shows the beginnings of an organised service in the Rhondda Fach, while Ystradyfodwg indicates that there was a post going well into the Rhondda Fawr. Possibly the Ystradyfodwg post can be associated with two other early letter-carriers whose names have been recorded, John Wilkes and Mari Cwm Iaen. Nothing is known of them other than the fact, preserved in popular memory, that they served the Rhondda Fawr between Dinas and Treherbert before 1858.

The 1855 report also records daily deliveries at Cymer (where a receiving house set up that year) and at Dinas, each of which cost £3 a year, obviously not a very onerous duty in either case. £5 a year was allowed for a delivery to Hafod and 6s a week, or £15 13s a year for deliveries in 'Ystrad Valley'. The much higher rate of pay suggests a longer and a more arduous duty than the deliveries in the lower part of the valley at Dinas, Cymer and Hafod.

The foot post from Pontypridd to Dinas was upgraded to a horse post on 7 March 1858, extended by means of a foot post to Treherbert. Four new sub-offices were opened at Gyfeilon, Eirw, Gellidawel and Treherbert.[17] The contract for the ride between Pontypridd and Dinas was awarded to 'Mr J. Williams, Post-office, Dinas'. Gyfeillon, Eirw and Gellidawel were all in the lower Rhondda and the sites of collieries which had been opened in the 1840s and 1850s; Treherbert was near the head of the Rhondda Fawr where steam coal had been proved a few years earlier.

The contractor for this ride, J. Williams, of the post office, Dinas must be John Williams, in his day a famous local character better known as Shôn Gwaun Adda. He should not be confused with the earlier Shôn y Gwehydd. He was certainly sub-postmaster at Dinas in 1864 but it is not clear whether he was also sub-postmaster in 1858 or merely lived at the post office.[18] The historian 'Morien' devotes a short chapter to him in his history of Pontypridd and the Rhondda.

> There lived at Dinas till recently, one John Williams, better known by the name "Shôn Gwaun Adda", from the place where his parents dwelt in the neighbourhood. He was born at Blaen-Cwrach, Glyn Neath, but came to Dinas when a lad with his parents ... In the early fifties, and down to the time a better postal service up the Rhondda was established, he was the only Postman between Pontypridd and Dinas, the then terminus of trade in the valley. He began as Postman on his own account, receiving a penny for each letter he delivered en route between Pontypridd and Dinas. Afterwards he came to receive a small salary from the Post Office to supplement his earnings. He speculated in a donkey and purchased a long low cart on two wheels, called "the Royal Mail," and in this he conveyed small parcels up and down ... In those days, the principal road along which Shôn's "Royal Mail" travelled was rough and uneven, and the deep ruts by the sides caused the cart to stick occasionally, so that it

---

[17] *The Cambrian*, 19 March 1858, gives the exact date. Handstamps for the four new sub-offices were issued on 9 September 1857 (Broomfield 1996, 80-81)
[18] Between 1854 and 1864 the births of several children are recorded in *The Cambrian* to 'Mr. Williams', Ishmael Williams and J. Williams, all described as being of the post office, Dinas. The relationship of John and Ishmael is not known, but possibly they were brothers; neither is it clear which of them was the sub-postmaster, although John was described as 'postmaster' on one occasion, in 1864. Ishmael died in 1891, aged 70

required the united efforts of both the donkey and Shôn
to get the cart again in motion ... [19]

Morien's account is based on folk memory and appears to be a little confused. It is not possible to make an exact match between the events recorded in postal records and Morien's narrative, and it may be that Morien has conflated Shôn y Gwehydd and Shôn Gwaun Adda. Nevertheless it is a valuable record which brings out clearly the state of the postal service in the valleys at the earliest stage of their development.

Although it dates from a slightly later period, the following extract from a letter published in the *Western Mail* in 1871 illustrates the difficulties that must equally have existed earlier when attempts were first made to establish a regular delivery in the Rhondda valleys:

> The streets are not named and the houses not numbered,
> and in a population of 18,000 ... it is not an easy matter
> to find out the right owner of many a letter. Generally the
> houses are crammed with lodgers, who at the time of
> delivery are at work, and their names cannot be
> ascertained. Yesterday morning I visited Pentre, and I
> may state that that very morning the postmaster had been
> around with a letter addressed to "Mr. John Jones (or
> Evans), Pentre, Near Pontypridd". Asking the name of a
> certain lodger in one of the houses, he was informed by
> the landlady that his name was "Dai Pentyrch". In a place
> like this, where there are so many John Joneses, it is a
> common thing that the same letter is opened by two or
> three persons ere it will find its right owner.[20]

---

[19] 'Morien' 1903, 177-8. The author continues with some examples of Shôn's ready wit, which may have been more amusing in the original Welsh than they are in his English translation

[20] *Western Mail* 24 October 1871

Official receivers/Postmasters of Eglwysilan/Newbridge/ Pontypridd

	Appointed	In office	Resigned
Thomas Morris	1830		
D. Davies		1836	
John A. Smith	1839		1841
D.N. Thomas	1841		1841
John Evans	1841		1842
Charles Bassett	1842		1874

Chapter 10

The Western valleys of Monmouthshire: the private posts (to 1839)

Four rivers – the Rhymney, Sirhowy, Ebbw Fawr and Ebbw Fach – rise on the southern slopes of the Brecon Beacons and then carve deep parallel valleys through the hills of north-west Monmouthshire in a generally southerly direction. The Ebbw Fach falls into the Ebbw Fawr at Aberbeeg and then the Sirhowy joins them lower down at Risca. The combined rivers flow into the sea at Newport. The Rhymney, instead of following its natural course towards Newport, does an abrupt turn at Machen and then flows south-west to enter the sea near Cardiff. The upper reaches of these four valleys form a distinctive region in terms of their character and their history and can best be treated together. This applies to both their industrial and their postal history.

Until the iron and coal industries started to develop in the 1770s, the area was thinly populated and remote – 'a rather cold, damp hilly place', as it has been described.[1] It was mostly contained within the medieval parishes of Bedwellty, Mynyddislwyn, Llanhilleth and Aberystruth. There were small centres of population around each of these churches, but nothing comparable even to pre-industrial Merthyr or Aberdare. The nearest market towns were Abergavenny and Newport. It has been estimated that in the mid 18th century the population of the entire Ebbw valley was about 4,000, most of whom lived in the southern reaches below Risca. Even in 1801, when several ironworks had already been established, the combined population of the four parishes in the upper Ebbw and Sirhowy valleys was only 3,986.

Life in the pre-industrial valleys was largely based on subsistence agriculture. It was an egalitarian and unassuming society with no resident gentry and only a few substantial farmers. The main land-

---

[1] Elliott 2004, 19

owners were the absentee Morgans of Tredegar House (near Newport) and the Hanburys of Pontypool. The coal reserves were worked on a modest scale for local consumption and an iron furnace was in existence in the Sirhowy valley as early as 1587.

The activities of the ironmasters of Merthyr and Dowlais had shown that the *blaenau* contained all the resources that were required for a successful coke-fuelled iron industry. This led to the establishment of a group of furnaces in the Western valleys to the east of Merthyr, located typically on the floor of the river valleys and close to the outcrop of the coal and iron measures. The first of these ironworks was at Sirhowy (1778), followed by Beaufort (1779), Ebbw Vale (1790), Nantyglo (1791), Tredegar (1800), the Union ironworks at Rhymney (1802), Coalbrookvale (1820), Blaina (1823), the Bute works at Rhymney (1824) and the Victoria works near Ebbw Vale (1837). Of these Ebbw Vale (which incorporated Sirhowy after 1818), Tredegar and Nantyglo emerged as the largest and most important. By 1823 the output of these three works was exceeded only by Cyfarthfa and Dowlais.

A name which features prominently in the development of the Monmouthshire iron industry – and of its postal service – is that of Homfray. Francis Homfray, a Staffordshire ironmaster, took a lease of Cyfarthfa ironworks at Merthyr in 1782. He soon gave this up but in 1784 he encouraged his sons, Jeremiah and Samuel, to start the Penydarren ironworks. Samuel established himself as the dominant partner at Penydarren and Jeremiah became one of the partners in the newly formed Ebbw Vale Iron Company in 1789. In 1791 he became the sole proprietor and a month later formed a new partnership with Harford, Partridge & Company of Bristol, bankers and charcoal ironmasters. The Harfords were Quakers, which was to have a marked influence on the way in which the company and the town were to develop. Following internal disagreements the Harfords bought Jeremiah out and acquired sole control. The family remained owners of Ebbw Vale until 1842 when the company was forced into bankruptcy. The business was then acquired by a company headed by Abraham Darby of Coalbrookdale, Shropshire.

Jeremiah's brother, Samuel Homfray, married a daughter of Sir Charles Morgan of Tredegar Park, a valuable connection with one of the leading county families of Wales. In 1800, with two partners, he leased a large area of land from Morgan on Bedwellty Common near the head of the Sirhowy Valley. Here he started to build a new ironworks which was named Tredegar in deference to Sir Charles Morgan. The town that grew up around the works took the name Tredegar Iron Works and it was not until the 1860s that the suffix finally fell out of use, the town becoming known simply as Tredegar. Samuel Homfray handed over the management of the works to his son, also named Samuel, who remained in control until 1853.

While the iron industry was still in the early stages of its development the coal mining industry had already become established as a major component of the local economy. This was facilitated by the opening of the Monmouthshire Canal in 1797 with its two branches, one to near Pontypool, the other to Crumlin in the Ebbw valley. Both branches were extended further northwards by means of tramroads. In the Sirhowy valley there was no canal, but the Sirhowy Railway of 1805 served the same purpose in carrying iron and coal to Newport. However, it was not simply the canal and the tramroads that stimulated the early development of the coal industry in the Western valleys. A major factor in its development was the exemption from coal duty which Newport, alone of all the south Wales ports, enjoyed until 1830. Newport's duty-free status together with the construction of the canal led to the proliferation of coal levels, and later of deep mines, throughout the Ebbw and Sirhowy valleys with the specific purpose of supplying the coal trade rather than feeding the furnaces.

The expansion of the iron and coal industries resulted in a corresponding growth in the population of the Western valleys. The parish of Bedwellty, which included the Sirhowy, Tredegar, Ebbw Vale and Rhymney ironworks, increased from 1,434 inhabitants in 1801 to 27,183 in 1851; Aberystruth (Blaina, Nantyglo) from 805 to 14,383 in the same period. In Bedwellty the intercensal period with the fastest rate of growth was 1801-11 (220 *per cent*) which represents the initial spurt following the establishment of the Tredegar and Union ironworks. In Aberystruth the fastest growth appeared in the decade 1811-21 (149 *per cent*) which is probably accounted for by the successful re-establishment of Nantyglo in 1811. In both parishes there was also a period of rapid growth in the 1830s and this corresponds to the period of fastest growth in the Monmouthshire iron industry as a whole when output increased by 143 *per cent* in the decade 1820-30 and by 86 *per cent* in the period 1825-35.

To accommodate this influx of population a number of new settlements sprang up. Thus were born the towns of Beaufort, Rhymney, Tredegar, Ebbw Vale and Blaina, all adjacent to the works. Brynmawr followed from about 1815. Unlike the other settlements it did not grow up around an ironworks but was developed by the Beaufort estate to increase the value of their property on the Breconshire/Monmouthshire border. Abertillery too was a late development, its growth starting in the 1850s.

In the first phase of industrial development houses had of necessity to be provided by the companies, at least for their key workers. Beaufort and Sirhowy each started as a few rows of houses put up by their respective ironmasters. At Tredegar, however, the original concept was more ambitious. Homfray seems to have laid out the town as a planned settlement from the outset, with the works, the workers' housing and the ironmaster's house and park all seen as an integrated development with the Circle in the centre of the town as its focus. Tredegar represents a rare example of integrated town planning in south Wales at this date and it quickly developed into a functional town with a range of private shops. By 1819 the company owned 254 small houses, but the planned central area had been completely built up and already privately promoted overspill and ribbon development were occurring.[2] At Ebbw Vale, on the other hand, the first company housing was built in 1790/91, and the company maintained a policy of building houses itself rather than granting building leases until as late as 1852.

For the same reason as company housing was provided so too was the company shop, or truck shop. However, after the initial years of a settlement different models of development emerged. At Ebbw Vale the Quaker owners ran a classic company town; they were highly paternalistic but generally regarded as good employers by the standards of the day. However, at Tredegar the Homfrays seem to have encouraged independent tradesmen to set up in business – or at least to

---

[2] Hilling 2003 argues that the central Circle was a feature of the town from the outset and not an accidental later development, as has been claimed. The only comparable planned industrial settlement in south Wales at this period is Morriston to the north of Swansea, which slightly pre-dates Tredegar

have tolerated them. In 1808 Samuel Homfray built a market hall and by the 1820s there were nearly 20 independent shopkeepers. There was still a company shop, but it had no monopoly. Similarly at Brynmawr, the Beaufort estate granted leases to shopkeepers from an early date (the first in 1827, well before the town's main period of growth) and in 1846 granted a lease with the specific requirement that the lessee laid out the property as a market place.

The differing approaches of the Homfrays at Tredegar and the Harfords at Ebbw Vale are reflected in the pace at which these two towns developed. By 1840 Tredegar was an established market town: it had a recognisable town centre with a wide range of shops and a developing middle class. As well as the company housing there was plenty of privately built housing which had been put up in a pretty haphazard way by private enterprise. But it was not until the 1860s that Ebbw Vale really took shape as a town. Until then there had been three separate settlements of Cwm, Victoria and Ebbw Vale with most of the population living in company houses and buying their goods at the company shop; there were virtually no private shops, no middle classes and no effective local government.

As was normal in the early phases of the industrial revolution, the managers also took up residence locally in comfortable houses standing in their own well-kept grounds, often screened by trees from the noise and glare of the furnaces. At Sirhowy Matthew Monkhouse and Richard Fothergill, who had acquired the concern in 1794, each built a 'mansion' (as their employees termed it) across the river from the works. Samuel Homfray included plans for a house and park in his development of Tredegar, but it was his son who actually built it in about 1817 when he took over management of the works. He named it Bedwellty House and lived there until the 1850s. It still survives in a public park close to the town centre. The Harfords, and later the Darbys, lived at Ebbw Vale House which was built in 1817 and until recently survived as a sports and social club. Perhaps the most notorious of all the ironmasters was Joseph Bailey who controlled Nantyglo from 1811 where he instituted a harsh and mean-minded regime. He was joined by his brother Crawshay Bailey (generally regarded as the inspiration for 'Cosher Bailey' in the ballad of that name). Joseph Bailey built the mansion of Ty Mawr on the opposite side of the valley to the works. Only the foundations now remain. He also built the famous defensive towers at Roundhouses Farm, a little higher up the slope, said to have been intended as a bolt-hole where he and his family could take refuge in the event of a rising among his workers – a prudent move on his part, in view of his reputation.

~~~~~~

Private posts

At least three private posts are know to have operated in the Ebbw and Sirhowy valleys – from Newport to Abercarn, from Newport up the Ebbw valley to Nantyglo, and from Newport up the Sirhowy valley to Tredegar. In addition there were branch posts that connected with the Nantyglo and Tredegar posts and private arrangements for sending mail on the Abergavenny to Merthyr mail coach. We are better informed about these services, especially the one to Tredegar, than about any other private post in south Wales thanks to three main sources of information. The first is a request made in 1816 by Benjamin Hall, the ironmaster of Abercarn,[3] to have the private Abercarn and Tredegar posts made official. His application, together with the report of the surveyor, Samuel Woodcock, on the feasibility of the scheme are preserved in Post Office Archives. Nothing came of it and in 1832 a similar request was made by John Moggridge, a coal owner who lived near Blackwood in the Sirhowy valley. His letters to Freeling (detailed almost to the point of being a rant) are full of information on the Tredegar post, albeit very one-sided, as is another letter which he persuaded one of his neighbours to write in his support. Again, the surveyor had to investigate and Rideout's report provides yet more information on both the Tredegar and the Nantyglo posts. These too are to be found in Post Office Archives. Finally there is Evan Powell's *History of Tredegar* which was published in 1902 but which contains material, presumably handed down by word of mouth, from a much earlier period in the town's history.

---

[3] He was also one of the M.P.s for Glamorgan and a partner in the Union ironworks at Rhymney. He is perhaps best known as the father of Sir Benjamin Hall, later the first Lord Llanover, who gave his name to Big Ben in the rebuilt Houses of Parliament

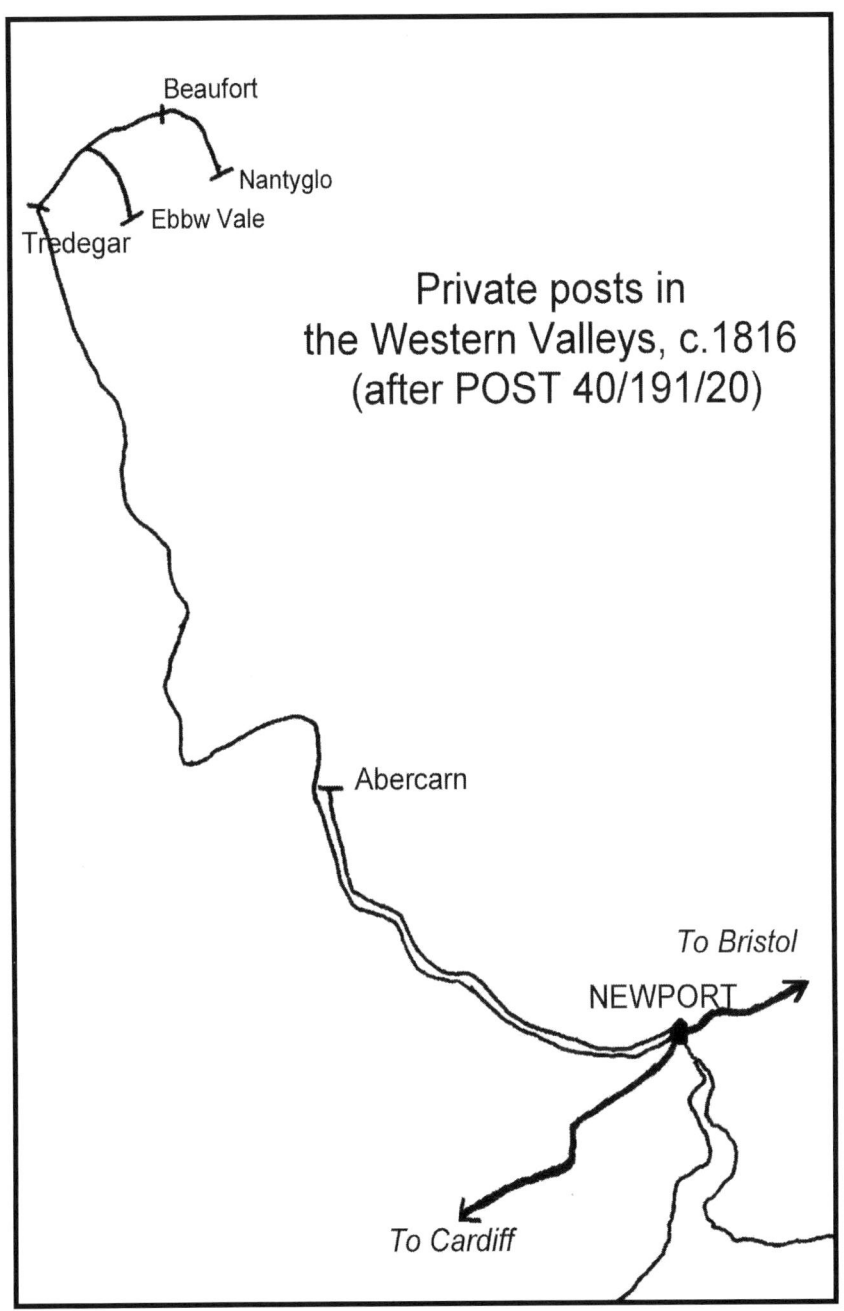

Private posts in
the Western Valleys, c.1816
(after POST 40/191/20)

The Tredegar private post

It is likely that the Tredegar post was established at much the same time as the ironworks, which was started in 1800. It is clear from Post Office and other sources that it was operated by Samuel Homfray of Tredegar but with the support of subscriptions from other companies such as Ebbw Vale and from a few private individuals. Powell indicates that it was in existence by about 1805/06: there seems to be no reason to doubt this date, since by then the Sirhowy Railway was open and there were three furnaces in blast. A postal service would have been essential. Powell writes:

> Postal accommodations were in a backward state,
> irregular and expensive; letters were conveyed from
> London at 1s. and 1s. 2d. each, left at the Company shop,
> and placed in the windows "until called for", or if
> delivered, 6d. extra.[4]

It would be unwise to read too much into the rates that Powell mentions. At the beginning of 1805 the charge for a single letter from London to Tredegar would probably have been 10d (8d from London to Newport plus a probable 2d for the private post). In March the General Post rate was increased to 9d giving a total charge of 11d. Powell's 1s 2d seems to represent the rate that was in force in the final years of the private post (10d plus 4d). The fact that letters were left at the company shop is significant, since it makes it clear that the service was just one more component of the infrastructure provided by the ironmaster. In its early days the post was probably carried by a postman on foot or – more likely – on horseback. In about 1829 a carriage road alongside the Sirhowy Railway replaced the primitive ridgeway which had previously been the only road up the valley, and once this road was available, and certainly by 1832, a mail cart started to be used..

In 1816 Benjamin Hall submitted his request for an official post from Newport via Abercarn to Tredegar. In his letter to Freeling he included a little sketch map which shows that the Nantyglo and Tredegar posts were operated as one at this date. The messenger from Newport went straight up the Sirhowy valley to Tredegar, by-passing Abercarn and

---

[4] Powell 1902, 30

Newbridge. A messenger continued from Tredegar to Beaufort and Nantyglo, with a branch from Beaufort to Ebbw Vale.

In his report on the proposal Woodcock noted:

> Tredegar, with Beaufort, Sirhowy, Ebbw Vale, Nantyglo
> ... This district is served by two Messengers, paid by
> subscription; one of them has wages per week 19/- and
> makes 5/- perquisites; the other has wages per week 17/-
> and makes 5/- perquisites ... These men collect and
> distribute letters.

Perhaps the messenger who earned 19s did the leg from Newport to Tredegar, the other the stretch from Tredegar to Beaufort and Nantyglo. The basic wage which these two men earned was certainly a lot more than they could have expected from the Post Office: at a slightly later date (1834) the messenger between Merthyr and Aberdare received 14s a week, and in 1843 the messenger from Bridgend to Maesteg received only 12s. The subscription element of their pay presumably came from the ironmasters and the few other contracted users of their services, the perquisites from fees paid for occasional collections and deliveries.

Woodcock concluded that whilst the number of letters was 'considerable', the income would not equal the cost of a Penny Post covering such a wide district. The total number of letters for Tredegar and the other ironworks was given as 20,352 per annum. At 1d per letter this would have yielded an income of £84 16s which would not have covered the costs of two messengers (even at lower wages) and of receivers at Tredegar and Nantyglo. But tellingly, Woodcock's report concluded 'and Mr. Homfray, the great Proprietor of the Ironworks, objects to any alteration'. Homfray had written to Freeling as soon as he had had wind of Hall's proposal. He claimed that the inhabitants of the neighbourhood were 'very well satisfied with the cheap charges' made by a private post that 'has long been very regular and answers the purpose well'. Furthermore, as 'Treasurers to the Establishment' Homfray & Co had invested heavily in the service and if it were to be replaced with an official post, they would have no way of recovering their money.

So far as the ironmasters were concerned Homfray was quite right: the ironmasters enjoyed preferential rates and the service was cheap for

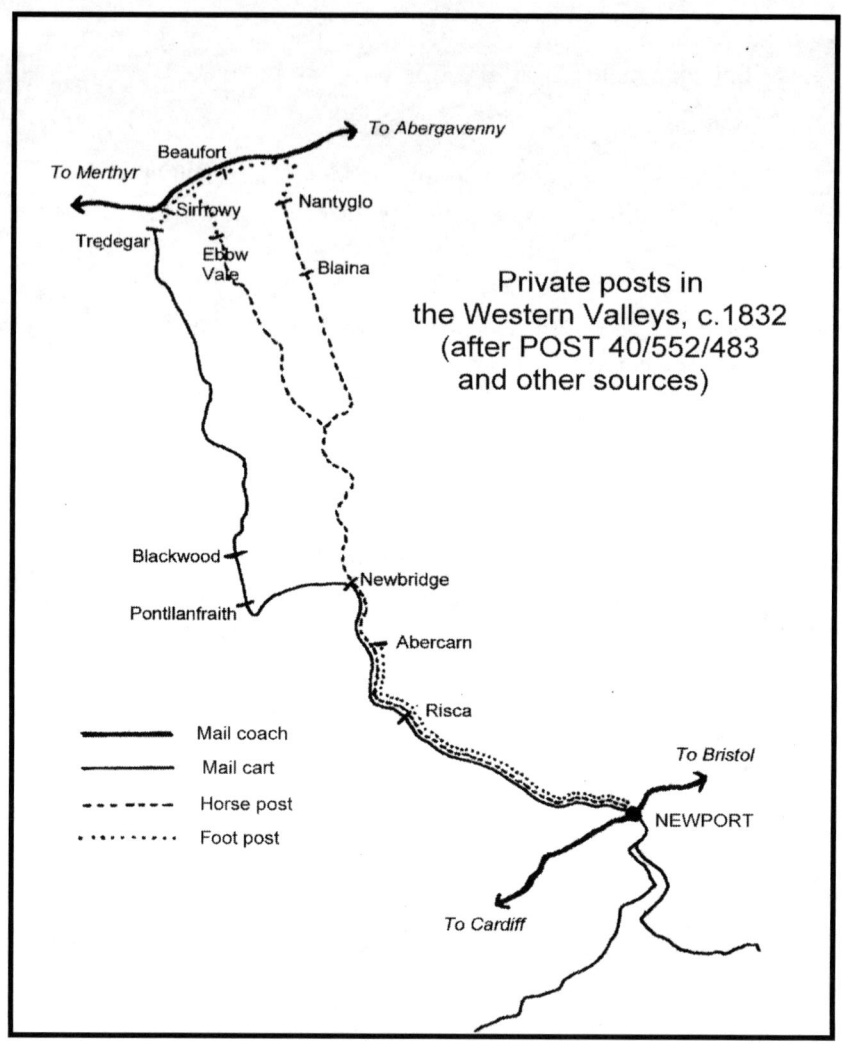

Private posts in
the Western Valleys, c.1832
(after POST 40/552/483
and other sources)

To Abergavenny
Beaufort
To Merthyr
Sirhowy
Nantyglo
Tredegar
Ebbw Vale
Blaina
Blackwood
Newbridge
Pontllanfraith
Abercarn
Risca
To Bristol
NEWPORT
To Cardiff

━━━━━ Mail coach
───── Mail cart
- - - - Horse post
· · · · · · Foot post

them, and it met their purposes well enough. But for the other inhabitants of the Ebbw and Sirhowy valleys the picture was not so rosy. Unless they were able to make their own arrangements to go to Newport with their letters they were obliged to depend on the ironmasters' post which for them was inconvenient and far more expensive than a Penny Post would have been. This came to light very clearly in 1832 when John Moggridge of Blackwood made a further attempt to have an official post introduced to the valleys.

Moggridge was an interesting character. His main base was in Cardiff but by the 1820s he had acquired a mineral estate in the Sirhowy valley some miles south of Tredegar. Here he developed a model community based on the principles of Robert Owen. By 1828 his community had 260 houses and 1,550 inhabitants, with shops, chapels, workshops, a school, a market house and medical provision. This was the origin of the town of Blackwood. Moggridge himself lived nearby in a house called Woodfield.[5]

By 1832, when Moggridge took up the issue of an official post, the Tredegar and Nantyglo posts were being operated independently of one another. A one-horse cart, driven at this date by a character called Charles Long, left Tredegar early in the morning and arrived in Newport at 10.00 a.m., in good time to connect with the up London mail. It returned at 3.00 p.m. after the arrival of the down mail and reached Tredegar at 7.00 p.m. The separate arrangements made by the ironmasters of Nantyglo and Ebbw Vale will be described below.

This was the situation described by Charles Rideout (who by this time had succeeded Woodcock as the Surveyor) in his report on Moggridge's proposals. It is fully borne out by Pigot's *Directory* for 1835. This lists a post office in Tredegar, and to the uninformed reader there is nothing to show that its status was any different from the official establishments in towns such as Monmouth or Newport. The postmaster is named as Richard Fothergill[6] who appears elsewhere in the directory as a linen draper, although in truth his establishment was probably the company shop. The only mail to be listed was to and from Newport: it went out at 6.30 a.m. and reached Newport at 10.30 a.m. In the opposite direction departure from Newport was at 2.30 p.m. and arrival in Tredegar at 7.30 p.m. There is an entry for the Royal Mail coach between Merthyr and Abergavenny, which had started to operate in 1824, but there is no suggestion that it carried mail.

For the Tredegar Iron Company postage was free: the costs were either absorbed in general expenses or more likely covered by the payments made by other users of the post. The other ironmasters who were served

---

[5] 'Utopia Britannica: British Utopian Experiments 1325-1945' (accessed 18 November 2009). Available from world wide web:
http://www.utopia-britannica.org.uk/pages/moggeridge.htm
[6] He can probably be identified with Richard Fothergill II (1789-1851), son of Richard Fothergill I (1758-1821) who was one of the original partners in the Tredegar company. See *Dictionary of Welsh Biography*, 266

by the Tredegar post paid a preferential rate of 1d per letter. But anybody else, including coal owners such as Moggridge, farmers, tradesmen and private families were hit hard if they wanted the convenience of a postal service without having to go to Newport for their letters. They had two choices: they could have a private bag and pay an annual subscription ranging from £5 to £7 to have it carried to and from Newport. It was not, however, delivered to their address, but put off at an agreed point, generally a cottage on the side of the road. The owner of the bag then had to make his own arrangements for its collection. Alternatively the user could pay on an item-by-item basis. In this case the letters were delivered to a private receiving house ('a self constituted office', as Moggridge termed it) from which he had to collect them. Such offices are known to have existed in Blackwood, Risca and Pontllanfraith. In 1832 the charge was 3d for each letter delivered from Newport, 2d for each letter sent to Newport and 2d for each newspaper, delivered or despatched. Apparently these charges had been substantially increased shortly before Moggridge wrote his initial letter to Freeling in April 1832. This suggests that the original charges might have been 2d and 1d respectively. By 1839 the rates had been increased once again to 4d per letter,[7] up or down. And as well as the cost, the local inhabitants also had cause to complain about the reliability of the service: Moggridge asserted that '[l]etters ... known to have been sent have never been received, and frequently do not get to hand until weeks after date.'

Not surprisingly the locals sought alternative means of getting their letters to and from the post office in Newport, short of actually making the long journey in person. An acquaintance of Moggridge who lived in the Sirhowy valley somewhere between Risca and Blackwood described his experiences in a letter to Freeling written in May 1832. Unfortunately his signature is illegible on the document preserved in Post Office Archives. Until April he had paid an annual subscription for a bag. He then decided that the time it took a servant to take it to or collect it from the receiving house could not be justified and he would make his own arrangements in future. At first he got hauliers on the Sirhowy Railway, which ran close to his house, to carry his letters to and from Newport, but before long he received a threatening visit from

---

[7] *The Cambrian* 8 February 1840

Charles Long, the mail-cart driver. Long, allegedly with the backing of Homfray, told him that it was a finable offence for hauliers on the railroad to carry letters. He was clearly wrong, but this unpleasant incident shows just how jealously Homfray guarded his monopoly. As a result of Long's visit the writer felt that he could no longer take advantage of that method. He was then reduced to sending a servant down to Newport or else having his letters sent to the receiving house at Blackwood and waiting several days for delivery by the 'humble individual' who kept it.

Rideout's report, endorsed by Freeling, came out against introducing an official post, and for very questionable reasons. He reckoned that the annual cost of a horse post from Newport to Tredegar with a branch from Newbridge to Nantyglo would be at least £280. In addition sub offices would be needed at Risca, Newbridge, Tredegar and Nantyglo bringing the total cost to £330. He estimated the likely revenue at rather under £450 and then came to the surprising conclusion that 'the Revenue would not warrant the expense'. Some might regard a return of over 33 *per cent* as definitely worth the outlay. He then put forward another dubious objection: if an official post were to be established it would not be possible to maintain the existing connection with the Bristol mail and so the shipping agents at Newport would lose a day in communicating with customers at Bristol. But if this connection was possible with the private post, why could an official post not be run to the same timings?

The truth must surely be that pressure was brought to bear on Rideout to ensure that he did not propose any changes to the existing system. He took the hint and concluded: 'In fact I do not think it advisable to interfere with the present arrangements, as ... the intermediate places now have the benefit of the Iron Masters private post'. Freeling saw no reason to question this conclusion. In his minute to the Postmaster General he wrote:

> ... further it would disturb the present arrangement which
> the great Iron Masters in that district have long adopted
> for their correspondence & with which it has been
> repeatedly decided this office ought not to interfere. The
> inconvenience felt by Mr. Moggridge is not denied ... but
> Mr. Rideout states that no Plan occurs to him, altho' he is

> perfectly acquainted with the Local circumstances, by
> which it can be accomplished at present.

The underlining appears to be Freeling's. In a word, he was simply afraid of doing anything that might upset Samuel Homfray. He was nearing the end of his career and no longer the man he had been thirty years before. He wanted to go out quietly rather than face a struggle.

Ebbw Vale and Nantyglo

As stated above, in 1816 the ironworks at Sirhowy, Ebbw Vale, Beaufort and Nantyglo were all served by an extension of the Tredegar private post which was operated by Samuel Homfray. This arrangement was still in force in 1827 when a copy letter to the Tredegar Iron Company appears in the Ebbw Vale letter book: 'We enclose a bill [i.e. cheque] for our Subscription to Sirhowy post to March 31st last ... '.[8] However, by 1832 separate arrangements were in existence for the iron companies in the Ebbw Fawr and Ebbw Fach valleys. In his report on Moggridge's application for the establishment of an official post in the Western valleys, Rideout reported that Ebbw Vale was served by a horse postman who left Ebbw Vale in the morning and arrived at Newport in time to connect with the up London mail and returned after the arrival of the down mail so as to reach Nantyglo at about 6.30 p.m. His sketch map shows that the main ride terminated at Nantyglo, but that there were two branches, one to Blaina and one to Ebbw Vale and Sirhowy (which had both been in the same ownership since 1818).

By now Ebbw Vale was no longer dependent solely on the private post from Newport. They maintained it for letters to and from Bristol, Newport and the south-west, which were all important destinations for them, but since 1824 their other mails had been routed through Abergavenny. In 1824 a mail coach had been started from Merthyr to Abergavenny. It was primarily designed to provide a faster and more reliable route to Merthyr for that place's London and North mails. Very soon after it started Ebbw Vale realised that they could take advantage of the coach to improve the speed of their correspondence with London, Ireland and the north. During September 1824 they notified several correspondents in these directions that in future letters should be sent via Abergavenny rather than Newport. Presumably they then

---

[8] GRO D.2472.2, 11 May 1827

arranged with the postmaster at Abergavenny that he would make up a bag for the company which would be carried on the coach (but as a private package, not as a mail). This would then be handed over to a representative of the company, probably at the *Rhyd-y-Blew Inn* in Beaufort which was the half-way point between Abergavenny and Merthyr at which the horses were changed.[9]

It is quite likely that Nantyglo made similar arrangements. A letter of 1827 from the office of Joseph and Crawshay Bailey and addressed to Usk, now in a private collection, was put into the post at Abergavenny. It may of course have been that a company servant was sent to Abergavenny with the letters, but in view of what is known of Ebbw Vale it seems likely that Nantyglo also had a private bag which was carried to and from Abergavenny on the coach. Further examples of mail from Nantyglo in the period 1824-39 would probably clarify the matter.

There is further evidence that the Abergavenny-Merthyr coach carried mail for places along its route on a private basis. Powell claims that letters and parcels were sent from Tredegar to Abergavenny by the coach[10] and in this he is supported by evidence from a slightly later date. In 1855 the Revd Thomas Rees, an Independent minister at Beaufort, complained, among other things, that Beaufort did not have a daily North mail:

> Owing to this inconvenience some of us are obliged to
> send our North letters to Abergavenny with the driver of
> the mail coach and pay him an extra penny for each
> letter, and we do this almost every other day ...[11]

However, this sort of thing was a purely private arrangement between the driver of the coach and the person with a letter to post. Since there was no official post to places such as Tredegar or Ebbw Vale the question of defrauding the Revenue did not arise and in the eyes of the Post Office it was just another private post.

The route of the coach took it through Tredegar, Sirhowy, Beaufort and Brynmawr but no attempt was made to serve these places, let alone

---

[9] Reynolds 2007 referring to GRO D.2472.2, 26 May 1827
[10] Powell 1902, 45
[11] POST 14/66

Rhymney, Ebbw Vale or Nantyglo which lay a little way off the road. An 1827 time bill for the coach shows no intermediate stage between Merthyr and Abergavenny and until 1839 there was not even one official receiving house along the entire 18 miles. Once more, the influence of the ironmasters had ensured that the Post Office did not adopt this obvious, simple and inexpensive improvement. At one time Rideout had considered the possibility of the mail coach stopping at Tredegar in conjunction with an official post from there to Newport. He tried to interest John Guest of Dowlais in the idea and Guest called a meeting of the ironmasters at Sirhowy, but not surprisingly the idea met with total rejection, orchestrated no doubt by Homfray.

Tredegar to Nantyglo

As originally established the private post between Newport and Tredegar continued from Tredegar to Nantyglo and this was the only way in which Ebbw Vale, Beaufort and Nantyglo received their letters. By 1832 Ebbw Vale and Nantyglo were being served more directly, but the post from Tredegar to Nantyglo was clearly still in existence. Rideout may not have referred to it in his 1832 report but Moggridge mentioned it and believed that the postman normally had five bags to deliver. Presumably these contained letters from Blackwood and Tredegar to Coalbrookvale, Sirhowy, Beaufort, Ebbw Vale and Nantyglo. Moggridge also referred to a grim incident that had taken place not long before he started to write his letters to Freeling, a full report of which appeared in the press:

> MURDER NEAR NANTYGLO. – While the postman, David Griffiths, was on his way from Tredegar to Nantyglo, Monmouthshire, he was overtaken by a shower of rain, which induced him to take shelter in a hut, where he saw an individual long known in the neighbourhood as a person of deranged intellect. The former and latter were seldom seen without pipes in their mouths. On the postman lighting his pipe, the other asked him for tobacco, which he said he could not afford; when the wretch struck him a violent blow with his fist, and proceeded to murder him with what they call a spanner, and [*sic*] instrument used for forcing a nut or screw. The unfortunate victim was immediately rendered insensible;

but, shewing symptoms of life the maniac again seized
the same weapon and actually beat his brains out. He then
left the hut, took off his shoes and hat, placed them on the
edge of a pond at a short distance, smeared himself all
over with clay, and ran off. He was apprehended at
Nantyglo. On being asked if he knew anything of the
murder, he answered "no"; but when shewn the body, he
said, "O yes (taking hold of the nose of the corpse), that's
the fellow who refused to give some tobacco!" The
postman has left a wife and seven children to deplore her
loss.[12]

Thomas Rees, the alleged murderer, was brought to trial at the
Monmouth Assizes in April. He was found guilty of manslaughter and
sentenced to transportation for life. In the press report of the trial the
name of his victim was now given as David Davies.[13]

Abercarn

In 1816 a private post was in existence from Newport in Abercarn,
about 10 miles up the Ebbw valley, and separate from the Tredegar
post. It must have been set up primarily for the benefit of Abercarn
ironworks, the property of Benjamin Hall, but it also served collieries
and ironworks lower down the valley including Tydu, Pontymister and
Risca. It was funded by the subscriptions paid by these concerns. The
messenger received £50 *per annum*, rather less than the Nantyglo and
Tredegar postmen received at this time.

It was Hall's suggestion in 1816 that this post, together with the
Tredegar and Nantyglo posts, should be converted into an official post.
Taken on its own, the financial case for an Abercarn Penny Post
appeared to be weak. Woodcock estimated that the number of letters in
a year was 6,120. At 1d per letter, his would have produced £25 10s
which would not have covered the cost of the messenger and the one or
two receiving houses which would have been needed. Similarly the
costs of an official post to Tredegar and Nantyglo would not have been

---

[12] *The Cambrian*, 24 December 1831. A shorter report had previously appeared in the
Monmouthshire Merlin 17 December 1831. It is interesting that the writer thought it
necessary to explain the function of a spanner
[13] *The Cambrian* 7 April 1832

covered by the likely receipts. Woodcock did not attempt to cost a combined post with the Tredegar post diverted to serve Abercarn instead of going straight up the Sirhowy valley. The financial case would have been stronger although it would still not have covered its costs, but in any case he would have received no thanks: Homfray had written a strong letter to Freeling opposing the idea and maintaining that everyone was perfectly happy with the existing arrangements and this carried the day.

The Abercarn private post continued, but it is not clear for how long. Moggridge makes no mention of it in 1832, but then there is no reason why he should have done so. Rideout's abortive proposals would have seen Tydu, Risca, Pontymister and Abercarn all served by a horse post from Newport to Tredegar, but since nothing came of this one can only assume that the private post continued until the establishment of an official post to Tredegar in 1839 which included a receiving house at Abercarn.

Rhymney

There is no documentary evidence at all from this period for the arrangements at Rhymney.[14] The Union ironworks of 1800 (also known as the Rhymney ironworks) must have had a postal service, and this was certainly quite separate from the Tredegar post. The Bute ironworks was built in 1824 by William Forman & Co of London. This company later purchased the Rhymney ironworks and ran the two concerns as one. No help is given by early directories, such as that of Pigot, since Rhymney was not regarded as being of sufficient importance to merit an entry. The only source of information that has been located are a few letters that have survived in private hands.

The first is a letter of 1822 from Rhymney to Chepstow. The markings indicate that it was carried by the Monmouth Penny Post from Crickhowell to Monmouth and then on to Chepstow by the General Post. The total postage was 9d (1d for the Penny Post plus 8d). It may initially have been carried by private hands over Mynydd Llangynidr to Crickhowell. Alternatively it might have been carried by a haulier on

---

[14] 'The Woodcock papers' 1992, no. 60, 14 lists private bags from Cardiff to 'Lanrumney' and 'Rumney'. These are almost certainly the hamlets to the east of Cardiff, formerly in Monmouthshire but now incorporated into the city

the Brinore Tramroad to Talybont-on-Usk and then found its way by one means or another to Crickhowell. It is not possible to say whether this letter is typical.

The second item is an entire of 1839 from Portmadoc to Rhymney which is addressed 'Rhymney Ironworks, via Abergavenny'. This strongly suggests that Rhymney followed the example of Ebbw Vale and Nantyglo in having their North mail (and probably their London mail as well) sent to Abergavenny and then on in a private bag carried by the mail coach from Abergavenny to Merthyr.

Rhymney's commercial links were all with Newport and there is very likely to have been a private post between the two places for the Bristol and Newport mails. So far no evidence of this has been found.

The end of the ironmasters' posts

Samuel Homfray may have nullified the attempt to establish an official post in 1832, but the private post had already become an anachronism by that date and the Post Office's failure to take effective action was indefensible. By 1835 Tredegar had matured into a fully functioning town and was no longer a mere appendage of the ironworks. It had a full range of independent tradesmen and a population of about 4,000. Down the valley Moggridge's Owenite colliery settlement of Blackwood had expanded to about 2,000. In the Ebbw Fach Brynmawr was expanding with the active encouragement of the Beaufort estate, its population including a fair number of independent tradesmen. Many in these new and growing towns would have had occasion to send letters and they justifiably felt that they were entitled at least to the same level of service as their neighbours in Merthyr or Abergavenny. In Nantyglo and Beaufort, strictly controlled by the Baileys, and in Ebbw Vale and Sirhowy, equally firmly if rather more benevolently controlled by the Harfords, the number of independent tradesmen was smaller and consequently the demand may have been less. Altogether, the population of the census registration districts which contained the iron towns in the upper Ebbw and Sirhowy valleys, i.e. Aberystruth, Tredegar and Rock Bedwellty in Monmouthshire together with Llangattock and Llanelly in Breconshire, almost doubled from 23,360 in 1831 to 45,385 in 1841. The fact that the Post Office was prepared to allow such a large population to depend on a private postal service which operated primarily for the advantage of the ironmasters showed up the inadequacies of its unreformed state. '[W]e have seen nothing yet to equal the want of management in the manufacturing districts of South Wales', wrote a Bristol journalist, and he went on to denounce the shortcomings of the postal provision in the valleys.[15]

By the mid 1830s the need for postal reform had become an issue that attracted much attention at both national and local levels, in Tredegar no less than anywhere else. Towards the end of 1836 a draper in the town, David Morris, initiated yet another attempt to bring about the end of the ironmasters' post. He succeeded in getting six Members of Parliament behind him and this must have helped to ensure that his representations received the attention of the Post Office. Rideout was instructed to report, but a minor incident in 1837, while he was still making his investigations, showed that Freeling's views on the iron-masters' posts were shared by Colonel Maberly, his successor as Secretary to the Post Office. A Mr Fleet complained at the charge that he had to pay to receive his weekly copy of the *Merthyr & Cardiff Chronicle*. Maberly retorted robustly that

> The present arrangements are perfectly legal and it would
> be quite out of the question to force the Ironmasters to
> give up their private bags for penny posts.

Rideout reported in March 1838. He proposed a post from Newport to Tredegar, with a branch to Blaina, Nantyglo and Brynmawr. Maberly was still anxious not to offend Homfray, although he accepted that the Post Office could no longer prevaricate over the issue of replacing the ironmasters' post. He commented to the Postmaster General:

> You will no doubt direct the Surveyor to survey the
> district in question, in person & to communicate with the
> principal parties interested. The extent & population of
> this large portion of country & the amount of the
> correspondence appears to demand that it should be
> placed under regular official regulations & I have no
> doubt but that the result will be beneficial to the Revenue
> as well as satisfactory to the Inhabitants, altho' some
> parties may possibly object to the change to their
> arrangements in the first instance.

---

[15] *Bristol Mercury* 13 April 1839

Rideout was instructed to visit the district once again and to report fully on the proposal. No doubt he was also instructed, so far as he could, to mollify Homfray and his allies. He presented a further report in February 1839, again recommending an official post to Tredegar and Brynmawr. His proposals were accepted and from that July the ironmasters' 4d post was at last replaced by an official Penny Post. The arrangements that were made for the official post and the way in which postal services developed in this district will be described in the following chapter.

The ironmasters appear to have accepted this arrangement without protest. By this time there was a real prospect of uniform penny post and so there was no benefit to them in keeping their private post. If it was going to cost the same to put a letter into the General Post at Tredegar or Brynmawr as it would at Newport, then there was simply no point in maintaining their own arrangements.

Appendix

Table of the revenue and number of letters from the district served by the Tredegar and Nantyglo private posts

[From Charles Rideout's report on the feasibility of replacing the private posts with an official post, May 1832 (POST 40/552/484)]

	Month revenue			Number of letters
Tee Dee	3			80
Risca		17	6	40
Pontymester	3		10	110
Abercarne	2	5	10	70
Newbridge		12	6	50
Blackwood	3	14	2	210
Maesruddin		10		40
Woodfield		10		20
Rock		18	4	50
	15	9	2	670
Tredegar	10	3	4	510
Beaufort	1	15		70
Colebrook Vale	1		2	40
	28	7	8	1290
Nanty glo	3	3	4	170
Ebbw Vale	3	1	8	160
Sirhowey	1	3	4	90
Blaina	1		10	80
	36	16	10	1790

Chapter 11

The Western valleys of Monmouthshire: the official post (from 1839)

Until 1839 Tredegar, Nantyglo and Ebbw Vale were connected to the General Post at Newport by means of an expensive private post operated by the ironmaster of Tredegar, Samuel Homfray. The Post Office, in the persons of both Francis Freeling and Colonel Maberly had for many years shown themselves reluctant to interfere with this monopoly, despite continued public dissatisfaction. However, by 1836 Maberly accepted that the post would have to made official, regardless of whether Homfray was offended by the loss of a useful source of income.

Instructions were given to the surveyor, Charles Rideout, to report on how this might be achieved. His first report was presented in March 1838, and on the basis of this he was instructed to visit the district again and prepare a detailed scheme. He presented his final proposals on 10 February 1839; these were duly approved on 3 March and implemented within a few months. Maberly was still conscious that opposition could be expected and noted:

> ... that as the whole arrangt. is local & of a peculiar
> character it should be placed under the superintendence
> of the surveyor and not under the mail coach department.
> In regards to the observation as to a part of the road to be
> travelled being private,[1] I do not apprehend any future
> difficulty as I feel convinced that the advantages to be
> derived from a regular official post will be too valuable
> to the parties interested, to suffer them to throw any
> obstacle in the way.

---

[1] Maberly probably had in mind the road up the Sirhowy valley constructed under the terms of the Sirhowy Railway Act, which might have been regarded as having the same status as the railway, i.e. as the property of the promoters

Tredegar and Brynmawr

Sub-offices were established at Tredegar and Brynmawr and opened on 10 July 1839.[2] Both came under the control of Newport and at both an allowance was made for local delivery. The first sub-postmaster at Tredegar was David Morris who had been active in campaigning for the replacement of the ironmasters' post, and at Brynmawr, William Williams.[3] Both were appointed on 1 April 1839 and both were drapers by trade.[4] Handstamps were issued to both places on 12 April 1839 in readiness for opening. Also included in Rideout's proposals was a Penny Post from Newport and handstamps were issued to the receiving houses at Risca, Abercarn, Newbridge and Blackwood on 17 April; the receivers at the two latter places were also given a delivery allowance.

Rideout's proposals envisaged that the Abergavenny-Merthyr mail coach would be used for the London and North mails to and from both Tredegar and Brynmawr. At Tredegar an allowance was made for a 'messenger to the mail road' to meet the coach and collect the London and North mails from it. The coach followed what is now the A4048 road through Sirhowy, skirting Tredegar, rather than going down into the town. At Brynmawr the road passed through the town centre and so the need for a special messenger did not arise. In 1844 the London and North mails from Abergavenny arrived at Brynmawr at 11.00 a.m. and were despatched at 1.00 p.m. At Tredegar the arrival would have been later and the despatch earlier, both at about midday. The up mail from Merthyr also included letters from Swansea and west Wales which had been carried on the Swansea-Merthyr coach. The London and North mails for the Penny Post offices of Risca, Abercarn, Newbridge and Blackwood were still routed through Newport.

For the Western mail from Newport, Bristol, etc Rideout envisaged an agreement with the proprietors of an existing stage coach between Newport and Tredegar. Some sort of an arrangement would also be made for meeting this coach at Newbridge and carrying the bags to and

---

[2] *The Cambrian* 6 July 1839
[3] For the life and subsequent career of Williams, see Edwards 1997
[4] In 1841 Morris's premises were described as 'adjoining the Tredegar Arms, with the principal front towards the market, and the second to Bridge-street' (*Bristol Mercury,* 4 September 1841), or just downhill from what is now The Circle. Williams' premises in Brynmawr were on Bailey Street in a building that later came to be known as the 'Old Post Office Inn' (Edwards 1997, xc)

from Brynmawr but he did not specify exactly what this arrangement might be. The agreement with the proprietors of the Newport coach did not come into effect until 23 November 1839[5] and there is no evidence as to the means that were adopted in the first few months to carry the Western mail from Newport to Brynmawr and Tredegar: possibly a temporary horse ride was established. The coach had been running since September 1837 and was operated by the Tredegar and Newport Joint-Stock Coach Company, a partnership of 25 individuals of whom twenty were based in Tredegar and the other five in the neighbouring townships of Sirhowy, Blackwood and Newbridge.

There was no direct post between Tredegar and Brynmawr: the Abergavenny mail coach called at both places, but only the London and North mails were carried: local mails must have been sent into Newport and back again. In 1846, when a request for a direct bag made, it was stated that there was no demand, since there were only about three letters a day between the two places.

On the same date, 23 November 1839, as the coach from Newport was adopted by the Post Office the promised link between Newbridge and Brynmawr also commenced. It took the form of a conveyance of some sort running on the Monmouthshire Canal Company's tramroad. The evidence for this is contained in a Parliamentary report of 1841[6] where an entry appears for the 'Bryn   and Newbridge Tram Road' (the gap after 'Bryn' is in the original). The Canal Company, despite its name, owned an extensive network of tramroads which were operated on the same principles as the canal, i.e. anyone was free to use them so long as they paid the tolls and observed the regulations. The role of the Company was simply to provide and maintain the infrastructure, not to operate the traffic. The mail coach on the tramroad would therefore have been operated by an independent contractor whose identity is not known. Passenger carriage by private contractors on the Canal Company's tramroads had started in 1822 (if not before) and by 1839 was probably a well established practice. This is an early example of mail being carried by rail and perhaps a unique instance of mail being carried on a horse-operated tramroad.

---

[5] 'Return of Stage-Coaches etc. used by the Post Office as Mails, not being so originally'

[6] 'A Return of the number, names, and wages of mail guards' 1841, 393

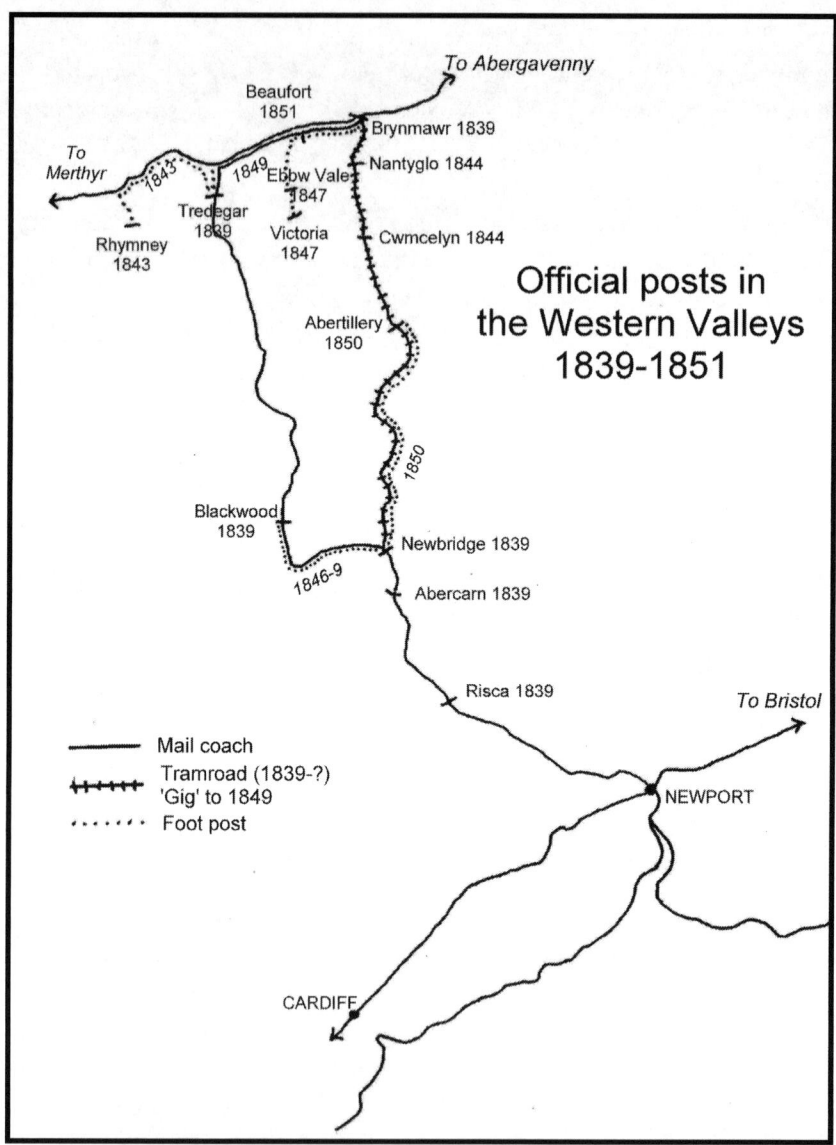

To Abergavenny

Beaufort
1851
Brynmawr 1839
To
Merthyr
1849 Nantyglo 1844
1843 Ebbw Vale
1847
Tredegar
1839 Victoria Cwmcelyn 1844
Rhymney 1847
1843

**Official posts in
the Western Valleys
1839-1851**

Abertillery
1850

1850

Blackwood
1839
Newbridge 1839
1846-9
Abercarn 1839

Risca 1839
To Bristol

NEWPORT

——— Mail coach
+++++ Tramroad (1839-?)
'Gig' to 1849
· · · · · · Foot post

CARDIFF

The tramroad actually extended all the way from Brynmawr to Newport, but below Crumlin it paralleled both the road and the canal and it was not necessary to extend the Brynmawr mail all the way. Instead it connected with the Tredegar coach at Newbridge. The timings make this clear. The Tredegar coach followed the modern A4048 down the Sirhowy valley as far as Pontllanfraith. There, instead of continuing down the A4048, it crossed to Newbridge in the Ebbw valley and continued to Newport on what is now the A467. At Newbridge it met the tramroad mail coming down the Ebbw valley from Brynmawr and the driver of the Brynmawr coach handed over his mail to the driver of the Tredegar coach (neither service had a Post Office guard). The two services came together thus:

Tredegar 7.30 a.m.	Brynmawr 6.15 a.m.
Newbridge 9.15 a.m. approx	Newbridge 8.15 a.m.
Newport 11.00 a.m.	

Newport 2.30 p.m.	
Newbridge 4.00 p.m. approx	Newbridge 4.15 p.m.
Tredegar 6.00 p.m.	Brynmawr 6.15 p.m.

(The times at Newbridge are the present author's estimates.)

It will be noted that the connection at Newbridge was much tighter on the northbound journey than on the southbound. Perhaps a generous margin was built in to the latter to allow for the possibility of delays caused by other traffic on a tramroad which carried a heavy traffic in coal and iron.[7]

These arrangements catered for the mails from Tredegar and Brynmawr to Newport. They also provided a fairly good connection at Newport with the mail coaches to and from London. These carried the Western mail from Bristol and the west of England for Tredegar and Brynmawr and all the mails for Risca, Abercarn, Newbridge and Blackwood. In 1840 the up London mail left Newport at 12.37 p.m. and the down mail arrived at 11.47 a.m. The exact times varied from year to year: as the service was constantly being speeded up, so the arrival time became earlier and the departure time later. By the beginning of 1843 the London mail was leaving Newport at 2.07 p.m. and arriving at 9.54 a.m.[8] Thus the Tredegar coach was now failing to connect with the west-bound mail coach, with the result that mail for Cardiff, Swansea and beyond had to be held in Newport for nearly twenty-four hours before it could be forwarded. In October 1843 the Abergavenny-Newport coach was re-timed to arrive at Newport one hour earlier in

---

[7] The timings for Brynmawr are taken from 'A Return of the number, names, and wages of mail guards'. H.C. 1841 xxvi 381, those for Tredegar from Pigot's *Directory* for 1842. See also Reynolds 1983

[8] *Monmouthshire Merlin* 14 January 1843

order to overcome this difficulty and, according to the *Monmouthshire Merlin*, 'we understand that negociations [*sic*] are now on foot to expedite the arrival of the Tredegar mail at this town' for the same reasons.[9] Nothing came of this, however, and the Tredegar coach continued to run to its existing timings until 1845.

At some date between 1841 and 1844 the Brynmawr service seems to have been transferred from the tramroad to the road, probably following the appointment of a new contractor. Pigot's *Directory* for 1844 has the following entry:-

> Letters from Newport, Bristol, &c. arrive (by mail gig) every evening at half-past six, and are despatched every morning at half-past six.

The use of the term 'mail gig' suggests a road vehicle. A gig is a two-wheeled, horse-drawn conveyance, basically of the same design as a mail cart. A mail gig would be allowed to carry passengers, whereas a mail cart usually did not do so; however, a mail cart was sometimes permitted to carry passengers and then, suitably adapted for the purpose, might come to be called a gig.[10] It seems unlikely that the term would be applied to a tramroad vehicle.

Changes to the Tredegar and Brynmawr mails

Major changes were made to mail distribution in south Wales on 1 September 1845. The London mail was re-routed via Gloucester, instead of Bristol, so as to take advantage of the recent opening to Gloucester of the Great Western Railway. This resulted in its arrival at Newport at 6.22 a.m. At the same time a stage coach already operating between Swansea and Bristol, the *Cymro*, was adopted to carry the North and Western mails between Bristol and south Wales. This coach arrived in Newport at about 12.15 p.m. (westbound) and 12.45 p.m. (eastbound). Thus there were now two separate mails on the main road through south Wales, the mail coach carrying the London mail and the stage coach carrying the North and Western mails. The Tredegar coach was re-timed to take account of this. The down journey now reached Newport at 5.20 p.m. to connect with the departure of the coach to

---

[9] *Monmouthshire Merlin* 7 October 1843
[10] For the distinction between a mail cart and a mail gig I am indebted to the late Revd Christopher Beaver

Gloucester and London at 6.38 p.m. The return journey left Newport at 7.30 a.m., an hour after the arrival of the overnight mail from London.[11] This took care of the London and Western mails reasonably well, but the North mail for the sub-offices between Risca and Blackwood was stranded at Newport for the best part of a day. The North mail for Tredegar and Brynmawr was still being received via Abergavenny.

This delay was clearly unsatisfactory and in July 1846 Rideout presented proposals for a revised service. It would seem that he had been instructed to examine the possibility of using the Taff Vale Railway between Cardiff and Merthyr and then using the coach to Abergavenny, and his report started by stating that this was not a feasible solution. It might conceivably have been workable for Tredegar and Brynmawr, but it could not do anything for the other sub-offices and so this attempt at saving money was discounted from the outset. Instead Rideout recommended accepting the tender of a local coaching firm, Lloyd, Walker & Co.[12] They proposed running a coach from Newport to Tredegar in the afternoon to meet the North mail, with a connecting coach from Newbridge to Brynmawr. For the London mail Lloyd & Co would run another coach in the morning as far as Newbridge with a 'single horse branch' (by which he may have meant a horse post or a one-horse cart) to Brynmawr. In addition the Post Office would need to provide a foot post from Newbridge to Blackwood. In commenting on Rideout's proposals, Maberly noted that the new arrangements would provide 'a new connection with the North mail, not at present enjoyed'. The total cost of coaching was put at £250 *p.a.* with a further 12s. per week for the Blackwood messenger.

These changes presumably took effect during the second half of 1846. The morning coach left Newport at 7.30 a.m. and carried London letters as far up as Newbridge (and on to Blackwood by means of the foot post) and local letters to Brynmawr. The afternoon coach (3.00 p.m. from Newport) carried the Western and North mails to all sub-offices as far as Tredegar with a connecting coach to Brynmawr. Places

---

[11] *Monmouthshire Merlin* 30 August 1845
[12] Thomas Walker may have been the landlord of the Parrott Inn at Newport who was a special constable at the time of the Chartist Rising of 1839. A Walker was also named as the driver of the mail coach from Tredegar to Newport which was involved in an accident in 1840 (*Bristol Mercury* 8 August 1840)

as far north as Blackwood now had two deliveries a day, the London and local letters in the morning and the North and Western letters in the afternoon. Tredegar and Brynmawr continued to receive their North and London letters from the Abergavenny coach. Brynmawr had three deliveries a day – the London, North, Western and overnight local mails at about 11.00 a.m., the Swansea and Merthyr mails at about 1.00 p.m. and the local day mail at about 6.30 p.m. Tredegar probably had two deliveries – at about midday for the London, North, Merthyr and overnight local mails and at about 6.30 p.m. for the local day mail. The local day mail understandably seems to have been only a light mail. In January 1848 it was proposed to discontinue the third delivery of the day at Brynmawr since there were only 21 letters a week for it. A year or two later Brynmawr was recorded as receiving 440 letters per week.

Come 1849 and the Monmouthshire Canal Company was in the throes of modernisation. As a result of much pressure from the ironmasters it had finally consented to convert itself from a canal company which also owned tramroads into a modern railway company which had a vestigial interest in a canal. The process was initiated by an Act of Parliament of 1845. A further Act followed in 1848 which, among other things, changed the name of the company to the Monmouthshire Railway & Canal Co and abolished the right of freighters to use their own horses on the line: henceforth traffic could only be worked by the M.R.C.Co's own locomotives. The date on which the new regulations were to come into force was fixed as 1 August 1849, but on that date only goods traffic was undertaken. It was intended to operate passenger trains, but the first service (from Newport to Blaina) did not start until 23 December 1850.

In May 1849 Rideout advertised – with a touch of desperation, one fancies – for a contractor to carry the mails to Tredegar and Brynmawr:

GENERAL POST OFFICE – NOTICE

Such persons as may be willing to undertake the conveyance of the mail, to and from Newport (Mon.) and Tredegar; Newport (Mon.) and Newbridge; Newbridge and Brynmawr; by a coach, on horseback, or in a close covered cart, proper for the purpose, are requested to send the terms upon which they will contract for the same, addressed to the Surveyor of the General Post Office, Gloucester, on or before the 1st June, 1849.

For particulars and forms of tender, apply at the Post Office at Newport, Brynmawr, or Tredegar.[13]

Although it is not stated in the advertisement, the new contractor was to take over from 1 August 1849. This was the date on which the M.R.C.Co was to become the sole operator of traffic on its system to the exclusion of private contractors. This suggests that the mail coaches operated by Lloyd & Co since 1846 might have run on the tramroad, and that with the impending changes to the regulations they had given notice to the Post Office that they could not continue after 31 July.

Clearly no tenders were received, and in July Rideout entered into negotiations with the M.R.C.Co to see whether they would be prepared to carry the mail from Newport to Blaina on their forthcoming passenger trains. The company was not prepared to make an offer 'in consequence of their arrangements for working the line being as yet incomplete', and with the cessation of the Tredegar coach now imminent, Rideout had no other course open to him but to recommend a fall-back solution – a temporary ride from Newport, through Risca and Abercarn, to Newbridge. This was duly approved on 24 July 1849 and came into effect on 1 August. The Blackwood messenger was not affected. Tredegar and Brynmawr still received their London and North mails from the Abergavenny-Merthyr coach. Their Western mail and local mail must now have been routed via Abergavenny (using the coach that ran through Pontypool).

The cessation of the Tredegar North mail coach was regarded as being only temporary, and Rideout still had hopes that the M.R.C.Co would soon be in a position to undertake the mail; but by November 1849 he had to accept that this would not be possible. Consequently he recommended that the separate North mail be formally discontinued and that the prospect of using the railway be abandoned. The best solution he could offer was a pair horse coach from Newport, running via Newbridge, Blackwood and Tredegar, to Brynmawr and this duly took effect on 26 November 1849.[14] The Blackwood messenger was

---

[13] *Monmouthshire Merlin* 12 May 1849
[14] *Monmouthshire Merlin* 23 November 1849

discontinued at the same time. In effect this service was the old Tredegar coach running to the timings of the Brynmawr coach and extended to Brynmawr. The new arrangements represented a reduction in the level of service, but were the best that could be offered. In particular, North mail items to or from Bristol and Swansea would be delayed for nearly a day by being held in Newport. The new service left Newport at 8.00 a.m. each day, after the arrival of the London mail. It carried London letters for offices up to Blackwood and local, North and West mail for all places en route. The return from Brynmawr was at 1.00 p.m. so as to be in Newport in time for the 5.42 p.m. despatch of the up London mail to Gloucester. Tredegar and Brynmawr continued to receive and despatch their London mail by the Abergavenny-Merthyr coach.

In June 1850 the South Wales Railway opened between Chepstow and Swansea: the following month the North and then the London mails were transferred to rail. The London mail was now routed through Bristol and across the New Passage once again. It arrived in Newport at 5.20 a.m. and the Tredegar/Brynmawr coach was re-timed to leave at 7.30 a.m. The following year the line between Gloucester and Chepstow was opened and on 19 July 1852 the Wye bridge at Chepstow was completed: at last there was a continuous railway line from London to Swansea. From 1 August 1852 the London mail was carried by rail throughout, now due in to Newport at 3.30 a.m. The Brynmawr coach was advanced once more to leave at 4.50 a.m. The departure from Newport in the evening of the London mail became correspondingly later, as did the return timings of the Brynmawr coach. It was probably at this point that the London mail to Tredegar and Brynmawr started to be routed through Newport.

The contractor for the revised 1849 service was Thomas Walker of Newport, presumably of the Lloyd, Walker & Co partnership who had taken the contract in 1846. He had originally undertaken the contract for £196 *p.a.*, but when it was re-advertised in March 1851 his tender had gone up to £224. Passenger trains between Newport and Blaina started in December 1850 and inevitably this had an effect on the viability of the road service. The Secretary to the Post Office clearly thought the price was too high, but was in no position to reject it:

> I submit that the lowest tender which has been received
> for the conveyance of mails between Newport and
> Brynmawr, viz that of Mr Thomas Walker who offers to
> perform the service by Pair Horse Omnibus or Car
> carrying passengers at the cost of £224 a year, may be
> accepted, but as the payment is in my opinion very high. I
> think it will be expedient to enter into a contract for one
> year only and to direct the Surveyor to advertise the
> service afresh at the expiration of that period.

When the surveyor did re-advertise in January 1853 the best price he could obtain for 'the Beaufort and Brynmawr Mail Cart Service' was £250 *p.a.* (This must refer to the mail cart from Newport to Brynmawr and Beaufort, not the mail cart between them.) In January 1854, at the time of Trollope's review, the mail cart was carrying direct bags to the sub-offices at Risca, Abercarn, Newbridge (including letters to Abertillery, Pontllanfraith and Gelligaer), Blackwood (including letters for Pontaberpengam), Tredegar (including letters for Sirhowy and Rhymney), Beaufort, Brynmawr (including letters for Ebbw Vale and Victoria, and for Nantyglo and Cwmcelyn).

The mail cart was still operating in 1858, apparently to the same general timings and following the same route, i.e. Newport-Tredegar-Brynmawr. It arrived at Tredegar at 8.00 a.m. and at Brynmawr at 9.15 a.m. The return journey left Brynmawr at 3.45 p.m. and Tredegar at 6.00 p.m.[15] Towards the end of 1858 part of the mail for Tredegar started to be sent via Merthyr on the Taff Vale Railway.

With the re-routing of the London mail for Tredegar and Brynmawr through Newport in 1852 and the consequent loss of payments by the Post Office, the Abergavenny-Merthyr coach was no longer a paying proposition for its proprietors. They therefore suggested to Rideout that instead of running in conjunction with the London mail at Abergavenny they should change their timings and connect with the North mail. They were even willing to do this without any payment from the Post Office, but simply for the privilege of exemption from tolls. Mail coaches were exempt from tolls and release from these heavy duties must have made

---

[15] These are the timings shown in Slater's *Directory* for 1858/59. The conveyance is specifically described as a mail cart in the listings for Brynmawr

the difference between profit and loss for the operators. The new arrangements started on 24 October 1852 – 'when the present London mail contract expires', as was noted in a Minute dated 4 October. The coach now ran from Merthyr to Abergavenny in the morning and back in the afternoon so as to connect with the up and down Gloucester-Llandovery coaches,[16] both due into Abergavenny at about 11.00 a.m. The result was that Tredegar and Brynmawr now received their North mail nearly a day earlier without it being held overnight at Newport. The Brynmawr letter carrier was allowed an extra 1s per week in wages, presumably because he now had to carry out a further delivery.

Initially the service on into England from Abergavenny was also by coach, but by 1854 a continuous line of railway existed from Chester to Newport and this was used for the mail. Timings between Merthyr and Abergavenny were altered as necessary to reflect the improvements brought about by the railways. Slater's *Directory* for 1858/59 specifically states that the mail from Abergavenny to Merthyr departs at 1.30 p.m. 'after arrival of train from Liverpool'. These arrangements for the North mail lasted until 1866[17] when the coach was discontinued. It was then replaced by a mail cart following the same route.

New sub-offices

Following the opening of Brynmawr and Tredegar in 1839 new sub-offices were opened and new delivery routes were created as the development of new centres of population and the increase in the volume of mail justified. However, throughout the period under consideration the service from Newport to Tredegar and Brynmawr, whether by coach or cart, formed the foundation for the circulation of mail in the Western valleys.

The first of the additional offices was at Rhymney. According to a contemporary newspaper report,[18] an office had been expected to open

---

[16] Following the opening of the South Wales Railway to Carmarthen in 1852 the Irish Mail was sent that far by rail and then on by road. The through coach from Gloucester to Carmarthen was discontinued and replaced by an abbreviated Gloucester-Llandovery service (cut back in turn to Gloucester-Brecon in 1853)

[17] This is the date given by Wilkins 1867, 352 which can probably be accepted since it was a very recent occurrence at the time of writing. However, the late Christopher Beaver (p.c. dated 5 May 2001) questions this, citing a circulation map of 1863 in the Royal Mail Archive which does not show Abergavenny-Merthyr as a mail route

[18] *The Cambrian* 6 July 1839

here in 1839 at the same time as Tredegar and Brynmawr, but this did not happen. When arrangements were being made for the opening of offices at Tredegar and Brynmawr, Messrs Andrew and Lewis, who were local tradesmen, requested a direct bag from Abergavenny to Rhymney.[19] This would have implied the opening of a receiving house. It was not an unreasonable request, since the mail coach from Abergavenny passed within a mile of the town but the request was turned down on the recommendation of the surveyor. His reasons are not recorded.

However, a few years later, in 1843, Rideout recommended that Rhymney should have an official post since more than 300 letters a week were now being received. Walter Goulstone was appointed Receiver on 23 November at a salary of £10 *p.a.* and a date stamp was issued a few days later. The office came under Abergavenny and opened early in 1844.[20]

Rideout's plan included a foot messenger who was to make three journeys a day to Tredegar and one to Rhymney Bridge, which was for the purpose of meeting the Abergavenny-Merthyr coach which carried the London and North mails. David Davies was appointed to this post on 23 December 1843. There was a free delivery within half a mile of the town centre which almost certainly was also carried out by the messenger. For all this he received 2s a day. It does not come as much of a surprise, therefore, to learn that in October 1844 the messenger was dismissed for his 'habitually irregular conduct', which, had he ever become aware of it, might have been the cause of momentary embarrassment to William Thompson, M.P., one of the partners in the Rhymney Iron Company, who had recommended Davies in the first place. A memorial was got up by the local inhabitants to have him reinstated but a local clergyman and the Receiver both supported the case that had been made against him and his dismissal stood. When it came to replacing Davies, the idea was floated that the Receiver should carry out the messenger's duties as well as his own, and this seems to have happened, for Walter Goulstone was appointed the Rhymney

---

[19] At a slightly later date Andrew, now in partnership with Buchan as maltsters, linen drapers, grocers and tea dealers, was one of the largest tradesmen in Rhymney (Scammell 1852)
[20] *Bristol Mercury* 3 February 1844

messenger on 22 February 1845. At the same time the number of journeys was reduced to two a day – one in the morning to meet the down Abergavenny-Merthyr mail coach in Tredegar, one in the afternoon to meet the coach at Rhymney Bridge as it returned. Within a few years, after suspicion had been cast on his conduct more than once, Goulstone was dismissed and prosecuted in 1850. It was then minuted that it was proving difficult to find anyone to undertake the duties at Rhymney 'on the present plan' (which presumably meant acting as both Receiver and messenger) and it was decided that Rhymney should be served by a messenger from Tredegar. This was still the practice in 1854 when Trollope listed the route as 'Sirhowy village and works, Dukestown, Rumney Inn and bridge, Rumney, Sodom, Gomorrah'. (Sodom and Gomorrah formed part of what is now Pontlottyn, allegedly taking their names from the harshness of life and the proliferation of brutal feuds in the village.)

In November 1844 an official post was authorised to Nantyglo, Coalbrookvale, Blaina and Cwmcelyn, since the weekly number of letters was about 569. They had previously received their mail through Brynmawr. Detailed instructions appeared in the press:

> All letters intended for Blaenavon [*sic*] and Cwm-Celyn, should be addressed to the Cwm-Celyn office, with no mention of Nantyglo or Brynmawr, otherwise they will be likely to be forwarded to either of those places. Letters intended for Garnfach and Nantyglo should be likewise directed to Nantyglo, without any mention of Brynmawr; and letters for Beaufort iron-works should be directed to Brynmawr, from which place they will be delivered.[21]

Receiving houses were set up at Nantyglo and at Cwmcelyn which was more or less equidistant from the ironworks of Blaina and Coalbrookvale. A date stamp was issued to Nantyglo on 19 December 1844 and Benjamin Williams was appointed Receiver on 3 January 1845. F. Hassie ('of Blaina') was appointed to Cwmcelyn on 31 December 1844, but it is not clear when this office actually opened, since the first recorded date stamp (incorrectly worded 'Cromcelyn') was not issued until 14 January 1846. Mail was delivered to both

---

[21] *Bristol Mercury* 28 December 1844

offices by a foot messenger from Brynmawr who received 12s a week, the same as his counterpart at Rhymney, but he seems to have had an easier time of it, for he was only required to do a single return journey between Brynmawr and Cwmcelyn, about three miles in each direction.

In 1847 sub-offices were opened at Ebbw Vale and nearby Victoria. Their mail was delivered by a messenger from Brynmawr. In 1853 his route was detailed as Beaufort, Beaufort ironworks, Ebbw Vale, Ebbw Vale works, Victoria, Victoria works. The Receiver at Victoria was allowed a salary of only £3 *p.a.* but his counterpart at Ebbw Vale was to receive £10 to cover the costs of delivery. Date stamps were issued to both offices on 10 August 1847 and a receiver was appointed at Victoria a few weeks later. But there was obviously some difficulty in finding a receiver at Ebbw Vale: in September the Surveyor was instructed to appoint, and again in November of that year. The problem lay in the requirement that the Receiver should also carry out a local delivery. No one was prepared to take on the commitment, which is not really surprising since the additional £7 salary represented a mere 2s 8d per week which would hardly have covered the wages of a letter carrier. In the end Rideout had to accept that the only way to find a Receiver for Ebbw Vale was to remove the local delivery requirement. On this basis a G. Thomas was appointed Receiver in January 1848 at a salary of £3 *p.a.*, the same as at Victoria, and the messenger from Brynmawr was instructed to carry out the delivery. Another reason, perhaps, for Rideout's difficulties lay in the nature of Ebbw Vale itself: unlike Tredegar or Brynmawr it was a company town. The company provided all the amenities and did not encourage private shopkeepers. Consequently, when the creation of an official receiving house was proposed, there were none of the independent drapers or grocers who in other places were available to take on the postal business as a natural extension to their existing activities.

A sub-office was set up at Abertillery in 1850, served by a messenger from Newbridge whose route included Crumlin, Llanhilleth and Aberbeeg (or 'Pontaberbeg' as Trollope had it). The first Receiver, Edward Jones, was appointed on 6 November 1850. On 3 April 1851 it was minuted that he was to be confirmed as Receiver 'on the understanding that his wife will personally attend to the duties'. This seems to be a case of a married woman running a post office, although to satisfy official requirements it had to be in her husband's name. A similar situation existed at Aberdare, Maesteg and (outside our area) Cowbridge.

A sub-office was opened at Beaufort in 1851: a direct bag was carried on the mail cart from Newport to Brynmawr and on from there by the Ebbw Vale messenger. Further sub-offices were established at Newtown (near Ebbw Vale), Sirhowy and Pontllanfraith (1854); Bedwellty, Crumlin and Tydee (i.e. Tydu) (1857); and Aberbeeg, Blaina, Pontlottyn and Pontymister (1858). A foot post was established in 1859 from Bedwellty to New Tredegar.

When they opened Tredegar and Brynmawr came under the control of Newport. By 1844 they had been transferred to Abergavenny. Abergavenny was thus responsible for all postal services to the west of the town nearly as far as Merthyr. In 1851 this process started to be reversed when Tredegar and Rhymney were transferred to Newport. In the following year a further batch of offices were similarly transferred, including Beaufort, Cwmcelyn, Ebbw Vale, Nantyglo and Victoria. Abertillery had been under Newport from the start. Finally, in 1858, Tredegar was made a post town and became responsible for a group of offices including Beaufort, Brynmawr, Ebbw Vale, Nantyglo, Newtown, Pontlottyn and Victoria. Newport retained control as far up the Sirhowy valley as Bedwellty and as far up the Ebbw Fach as Blaina.

Receivers and messengers

The Post Office seems to have experienced great difficulty in making suitable appointments of Receivers and messengers. Receivers rarely lasted more than a few years, messengers often less. Frequently a note appears in the official Minutes that the Surveyor is to select a suitable person for Receiver or messenger at a particular place, which indicates that attempts to fill the vacancy in the normal way had been unsuccessful. And once an appointment had been made there was no telling what he or she might not get up to: the Minutes are littered with reports of greater or lesser misdemeanours on the part of Receivers and messengers in this part of the world.

The resignation and prosecution of Walter Goulstone at Rhymney has already been mentioned. He was appointed Receiver in 1843 and in

February 1845 took on the additional role of letter carrier. Very soon afterwards he fell under suspicion. Maberly noted on 2 April:

> I cannot but concur with Mr Rideout that the explanations made by the Receiver at Rhymney as to the delay & opening of the letters from Mr Lindsay & Mr Morgan are very unsatisfactory & suspicious. Still as there is no direct evidence of his having detained or opened the letters wilfully, or from any improper motive, I can only recommend that he may be very severely reprimanded ...

No doubt he was, but it seems to have had little effect. On 28 April 1846 it was noted that £2 0s 3d had been lost from a bag sent from Rhymney to Merthyr. No explanation was offered, and the matter was not taken further, but in view of the previous incident Goulstone's suitability for his combined post must have been called into question. His downfall came a few years later: the Minutes do not go into detail, but by May 1850 he had resigned and Maberly was taking legal advice. Goulstone was duly prosecuted and convicted in June.

The premises in Bailey Street, Brynmawr occupied by William Williams when he was the town's first postmaster, 1839-43.

Whilst Goulstone probably deserved whatever sentence he received, sometimes aspersions were cast on a Receiver that appear to have been motivated simply by malice. In September 1840 Rideout had to investigate complaints against William Williams at Brynmawr. Again, no detail is given of the exact nature of the complaints but Rideout decided that it was nothing more than 'a malicious attempt to bring unfounded charges against the Sub Postmaster at Brynmawr.'

Williams resigned in 1843. He was replaced by Richard Marsden, a draper, who was appointed on 30 May. No attempt seems to have been made to examine his suitability for the post, for within a month it was discovered that he was an uncertificated bankrupt and so unable to give the necessary security for a stamp license. He was removed from office forthwith. He was succeeded briefly by David Edwards, a close friend of William Williams and the brother of his wife's first husband. He combined being a Calvinistic Methodist minister with a grocery and drapery business. According to a later writer he was 'for many years the most prominent public character in the district – full of energy, resources and method'.[22] He resigned within a year because he was unwilling to work on a Sunday[23] and was succeeded by Thomas Davies. He is perhaps the Thomas Davies who was listed as a grocer in Pigot's 1844 *Directory* and is the same as the 20-year old Thomas Davies who in the 1841 census described himself as a 'draper' but in reality was probably just a live-in assistant in a draper's shop.[24] In March 1846 the office was broken into and robbed, but Davies was still in office in 1858.

The behaviour of letter carriers was a constant cause for concern on the part of the authorities. Given the independent nature of their duties, away from any immediate supervision for most of their time, the temptation of the familiar human weaknesses was less easy to resist than it might have been in a more controlled environment. One of the

---

[22] A paper by John Thomas, read at a meeting of the Brynmawr Mutual Improvement Society, 30 November 1905 and published in *Tong's Illustrated Almanack*, 1906. Extracts available from: http://www.brynmawrscene.net [accessed 23 November 2009]. See too Edwards 1997

[23] *Bristol Mercury* 28 December 1844

[24] Information from Pigot 1844 and the 1841 census returns taken from Jeffrey L. Thomas, 'The Brynmawr Wales History – Genealogy Project'. Available from: www.thomasgenweb.com/brynmawr.html [accessed 23 November 2009]

first recorded offenders was Thomas Brockett, the messenger who had been appointed in 1846 to carry the mail from Newbridge to Blackwood. In October 1849, when rather the worse for drink, he lost his letter bag:

CAUTION TO LETTER CARRIERS. – On Saturday last, Thomas Brockett, post office messenger between Blackwood and Newbridge, was convicted before the magistrates, at their office, High-street, [Newport] on the information of Mr. Charles Rideout, surveyor of post offices in Gloucestershire and Monmouthshire, of having lost the letter bag on his road between the towns abovenamed. The penalty in this case is not more than £20, or three months imprisonment; but it being shown that Brockett was of unimpeachable character and general sobriety, although he had on the present occasion taken a little too much drink on the occasion of the marriage of the postmaster's daughter, the magistrates properly censured the conduct of the defendant, which they did not believe would occur again, and fined him 20s. and costs. We understand that all the witnesses who were brought down against Brockett, so esteemed the almost heart-broken and fully penitent man, that they would not accept one farthing for their expenses.[25]

As well as the fine imposed by the magistrate Brockett also incurred a penalty imposed by the Post Office, probably loss of pay. Almost immediately after this incident the Blackwood foot post was discontinued as part of the re-ordering of the mail services in the Western valleys in November 1849. Brockett was temporarily out of a job, but in October 1850, all forgotten and forgiven or perhaps because nobody else offered, he was appointed messenger from Newbridge to the newly opened office at Abertillery. He soon got into trouble again and in January 1852 he was stopped one week's pay 'as a punishment for his misconduct', but not dismissed,. The nature of this misconduct was not disclosed, but perhaps the postmaster had married off another daughter.

---

[25] *Monmouthshire Merlin* 17 October 1849

Drunkenness was certainly the cause of the downfall of Matthew Burge. He was the messenger from Brynmawr to Ebbw Vale and Victoria who had been appointed in August 1847. In January 1850 he was dismissed because he 'appears to be a confirmed drunkard'. Another incident at Brynmawr in 1852 concerned the letter carrier's wife who was allowed to resume her duty in accordance with the wishes of the principal inhabitants of the place. One can only imagine that she acted as a kind of volunteer postwoman to help her husband, and officialdom took exception to this.

Tredegar had its share of troubles, too. In 1846 the sub-deputy, John Lewis, was given a severe reprimand for irregularities in the transmission of public money: this sounds more like laziness or carelessness than anything criminal. The next complaint about Lewis also suggests that he tended to sit lightly to the rules. An anonymous complaint was made in January 1849 that letters had been destroyed at the post office. On investigation it appeared that this was not so, but it did turn out that Lewis had been in the habit of allowing his son to assist in the office, and that the son had never taken the official declaration. Lewis was warned that this must not be allowed to continue.

Very soon after this Lewis died and his widow, Mary Lewis, was appointed to replace him. She seems to have taken after her late husband, for in January 1851 she was given a severe reprimand for frequent neglect of duty as regards her remittances. She then became a defaulter again although her accounts were subsequently put into a satisfactory state, which shows that her fault was more one of inadequate book-keeping than of dishonesty. Rideout carried out an special investigation and produced a very unfavourable report, but the President of the Money Order Office stated that the money order business at Tredegar had been carried out in an improved manner (Tredegar had been an M.O.O since 1845) and the accounts were now prepared with 'an average degree of accuracy'. Maberly brusquely advised the Postmaster General: 'Give her a warning & a further trial, but let her understand it will be the last.'

However, there was one postmaster at least in the valleys who was an ornament to his community and to the cultural life of south Wales. This was Edward Evans, the postmaster of Rhymney, who merits an entry in the *Dictionary of Welsh Biography*, a distinction shared with Charles

Wilkins of Merthyr. Evans was born in Llanidloes, but his family moved to Rhymney in 1825 when he was two years old, part of the great migration from rural Wales to the valleys. In his youth he worked a miner, but by his own efforts he become a competent musician and writer. He opened a bookshop in Rhymney and was appointed postmaster. He was a distinguished and successful conductor and choirmaster until his death in 1878.[26]

Receivers/Subpostmasters at the principal offices

	Appointed	In office	Resigned
Brynmawr			
William Williams	1839		1843
Richard Marsden	1843		1843
David Edwards	1843		1844
Thomas Davies	1845	1858	
Rhymney			
Walter Goulstone	1843		1850
Edward Evans		1858	1878
Tredegar			
David William Morris	1839		1844
John Lewis	1844		1849
Mary Lewis	1849	1852	
Henry Joseph Crowe		1858	
Ebbw Vale			
G. Thomas	1848		
Abertillery			
Edward Jones	1850		

---

[26] *Dictionary of Welsh Biography* 1959, 228-9

Chapter 12

Pontypool and the Eastern valley (Monmouthshire)

Pontypool is one of the oldest industrial centres in south Wales. It does not actually lie within the coalfield, although the coal and the ironstone measures outcrop just to the west of the town, and could easily be worked with the technology that was available in the Middle Ages. The Afon Lwyd provided a steady supply of water, and limestone was readily available. It was thus a suitable location for early iron working, and a forge is known to have been in existence by 1425. The Hanbury family, who were largely responsible for the development and success of the town for over 300 years, first appeared in Elizabethan times, when Richard Hanbury set up an iron furnace. The first member of the family to live in Pontypool was Major John Hanbury, who took over control of the family industries in about 1685. He was responsible for the building of Pontypool Park, the house from which he and his family dominated the economic and social life of the town, both as squires and as industrialists, until they moved away in 1908.

Under John Hanbury and his successors the ironworks developed into one of the most important in Europe in the eighteenth century. It also saw the first successful manufacture of tinplate on a commercial scale in the British Isles. The production of sheet metal for tinplating led in turn to the japanning industry that is particularly associated with the town. Pontypool ware (lacquered and decorated sheet metal) was produced in the form of a great variety of decorative and domestic objects, which commanded a wide market among the better off in the eighteenth century and are still sought after today by connoisseurs. Japanning remained the preserve of one family, the Allgoods, and it flourished in Pontypool until the beginning of the nineteenth century. For various reasons the industry then migrated to the west Midlands, but the name 'Pontypool Japan' still survived.

The tinplate trade itself continued as an important component of the local economy throughout the nineteenth century at Pontymoile and in the area the south of the town that was later developed as the new town of Cwmbran. There was also a significant coal industry around Blaenavon and in the coalfield to the west of Pontypool. Initially, as in the rest of the Valleys, output was principally confined to furnace consumption and local use, but with the opening of the Monmouthshire Canal in 1796 a lively export trade through Newport developed.

Despite its location within the valleys, eighteenth-century Pontypool had more in common with coastal and county towns such as Swansea, Brecon or Abergavenny than with isolated upland villages such as Merthyr Tydfil or Aberdare. The local presence of a landed family with claims to county status and a well grounded economy with a steady market for its products led to the development of a stable and settled community well in advance of the beginnings of urban growth in the *blaenau*. This is shown in the provision of a market house as early as 1730, and by the short-lived existence in the town between 1740 and 1742 of a printer, Samuel Mason. The first turnpike road was built under an Act of Parliament of 1758, well before Merthyr or any of the other valley communities enjoyed similar facilities.

The ironworks at Pontypool remained in the hands of the Hanbury family until 1851. They were converted to coke firing in about 1807 and continued to specialise in the production of sheets for the tinplate trade and of wire. However, they declined in relative importance with the establishment of rival undertakings nearby, above all Blaenavon (1789), but also Varteg (1803), Pentwyn (1825), Abersychan (1825) and Golynos (1837). In 1851 Capel Hanbury Leigh sold the ironworks, thus ending the involvement of the Hanbury family in the iron industry that had lasted for nearly 300 years. The family remained at Pontypool Park until 1908, although no longer directly connected with the town's industries.

The town continued to thrive, both through the continuing presence of the iron industry and as the commercial centre of the Eastern Valley, with its iron, tinplate and coal towns. The handsome Town Hall of 1854-6, erected under the patronage of Capel Hanbury Leigh to commemorate the birth of an heir, is a clear indication of the growing status of Pontypool and an example of the benefits that it derived from the presence of a landed family who took a close interest in the town.

In the 1790s the population of Pontypool was estimated at about 1,500, which is fully consistent with the 1,472 recorded in the census of 1801 for Trevethin, the parish which included Pontypool, part of Blaenavon and the valley in between. This made it similar in size to Aberdare, but very much smaller than Merthyr. The population of Trevethin grew rapidly and was nearly 17,000 in 1851. There was a period of particularly strong population growth between 1821 and 1831 due to major developments in the local iron industry. Two completely new works, Pentwyn and Abersychan, were set up in 1825, and the capacity of Varteg was more than doubled between 1824 and 1830.

The Penny Post

The first mention of any sort of postal service comes in 1807, when the surveyor, Samuel Woodcock, reported that the town was served by a private messenger who carried the mail to and from Newport and charged 2d per letter. Almost certainly this private messenger service must have dated back well into the eighteenth century. It would have been required by the presence of a resident gentry family, and the existence of a successful industrial base, with its need to maintain contact with customers and agents.

Woodcock's report was occasioned by a request made by Sir Charles Morgan of Tredegar Park, one of Monmouthshire's M.Ps. and a leading figure in county circles, for a post from Newport to Abergavenny via Caerleon and Pontypool. Although Morgan's application has not survived and Woodcock does not state it in his report, there were probably three reasons for the request. One must have been to reduce the cost of postage by making Pontypool a post town and cutting out the 2d fee charged by the private messenger. A second reason was probably to have the London and North mails to and from Pontypool routed via Abergavenny so as to avoid the delays that were caused by the passage of the Severn. Situated as it was almost exactly at the mid point between Abergavenny and Newport, Pontypool could potentially receive its London mail from either place as quickly and cheaply as from the other. A third reason behind Morgan's request, judging by the arguments put forward in a similar application of 1814, was to improve the circulation of mail from Brecon and Abergavenny to Newport. As it was, such letters were routed via Gloucester and Bristol and took six days: a direct mail from Abergavenny to Newport would have reduced this to a single day.

Woodcock reported that the cost of providing a post between Newport and Abergavenny would be greater than the revenue that could be expected from the letters that might be carried. 'It might indeed be desirable to make a regular Post from Newport to Pontypool, which place has hitherto been served by a private messenger', noted Francis Freeling, but he could only recommend a Fifth Clause Post, i.e. a post guaranteed against loss by the local residents. This would also have required the residents to pay an additional fee for the collection and delivery of their letters which was not at all the idea they had in mind. They would surely much have preferred Pontypool to be made a post town. Rates would then be calculated from Pontypool, and given its closeness to Newport, in the vast majority of cases these rates would be the same as from Newport. In other words they would have delivery to Pontypool for the same price as they were paying for delivery to Newport and no need to pay the private messenger's fee on top. If the best the General Post Office could offer was a Fifth Clause Post, they would be no better off, and so it is quite understandable that this option was rejected.

A daily horse post between Raglan and Usk had been in existence for a number of years, connecting with the London to Milford Haven mail. In July 1814 Lord Arthur Somerset proposed that this post should be extended to Newport so establishing a link between Newport and the north of the county. Lord Arthur was a younger brother of the 6th Duke of Beaufort and M.P. for Monmouthshire from 1805 to 1816. It was a more modest proposal than that of Morgan in 1807, but it was designed to serve the same purpose.

Woodcock again presented a report. He noted that there was a considerable correspondence between Newport and Abergavenny which would benefit from the proposed extension, and further that links between Breconshire and Monmouthshire in general and Glamorgan would be improved. He proposed a three-day extension of the Raglan-Usk post at a cost of £33 *p.a.* but he also entered few caveats. Mail

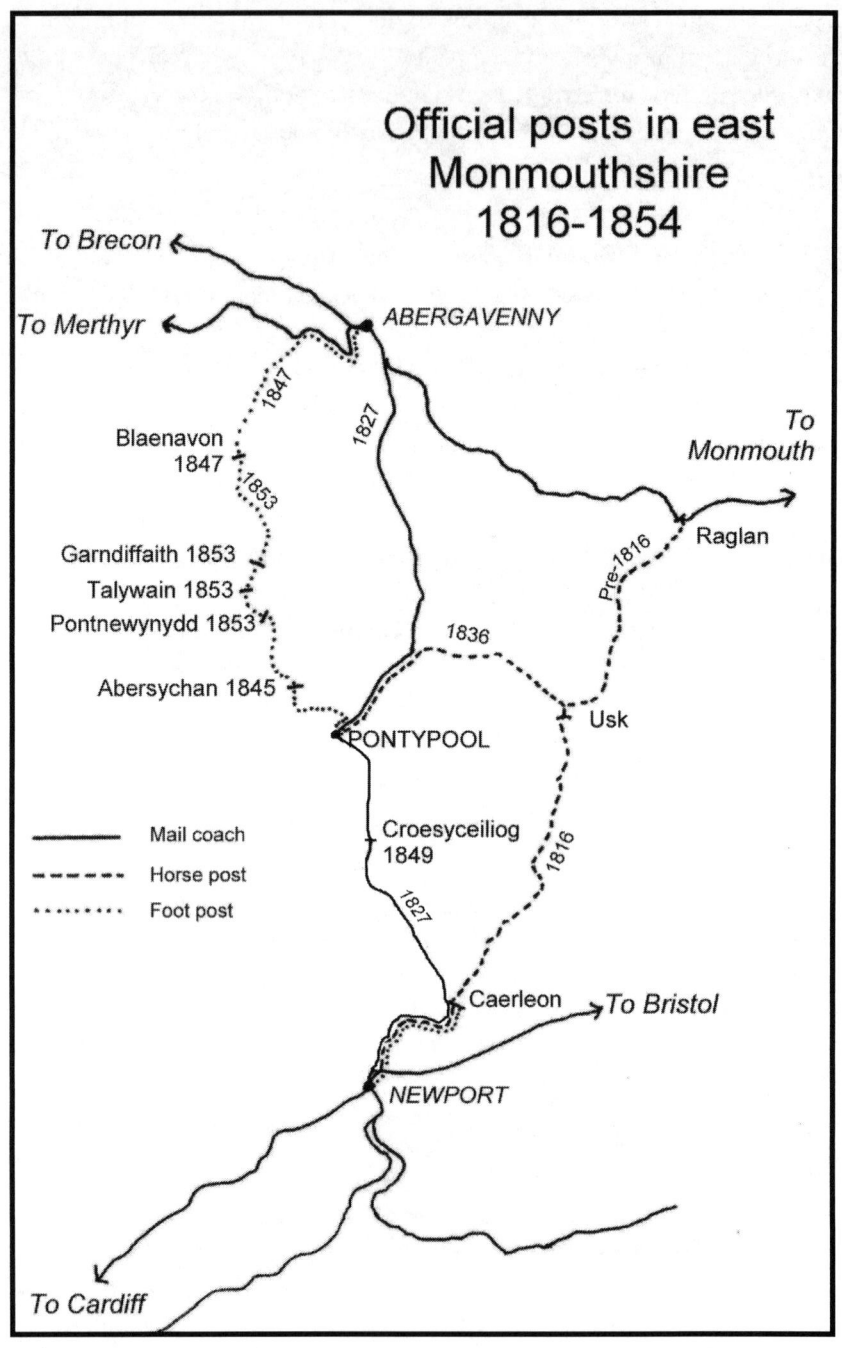

Official posts in east
Monmouthshire
1816-1854

To Brecon

To Merthyr

ABERGAVENNY

1847

1827

To
Monmouth

Blaenavon
1847

1853

Garndiffaith 1853

Talywain 1853

Pontnewynydd 1853

Pre-1816

Raglan

Abersychan 1845

1836

PONTYPOOL

Usk

Croesyceiliog
1849

1816

Mail coach

Horse post

Foot post

1827

Caerleon

To Bristol

NEWPORT

To Cardiff

between Abergavenny and Bristol would not benefit because the arrival of the Raglan post at Newport would be too late to connect with the coach from Swansea to Bristol, but above all the new arrangements would do nothing for Pontypool which lay well off the Raglan-Usk-Newport route:

> But Pontypool is not so well off; the trade produced by
> the Coal mines and Iron Works in the neighbourhood of
> that place is very considerable: from a return sent me by
> the Postmaster of Newport, (from whence they are
> served) it appears that the letters and newspapers to &
> from number more than 400 in one week. At present they
> have no Post establishment, but receive their letters by a
> Messenger from Newport, to whom, I am informed, they
> pay 2d. a letter.

If the Usk-Newport post were to be routed through Pontypool it would result in a lengthier journey time and missed connections at Newport to London and the south of England. Woodcock therefore proposed that a Penny Post be set up at Pontypool as a separate undertaking from the Usk-Newport post, but one which would still be profitable.

The Usk-Newport post was approved immediately but a full report on the proposed Pontypool Penny Post was called for. This must have proved satisfactory, for a Penny Post under Newport was in operation by 1816 which is the date of the first recorded mark.[1] In later years Pontypool also had a Penny Post under Abergavenny, but it is not clear whether this was set up at the same time as the Newport post or subsequently. There is some evidence to suggest that it was in existence by 1823. Complaints were made that year about delays in the service between Caerleon and Pontypool. In response it was stated that since neither was a post town, letters had to be sent to their principal offices, which were named as Newport and Abergavenny. Caerleon was described as a sub-office and Pontypool as a Penny Post. Since Newport was obviously the post town for Caerleon, this implies that Abergavenny was responsible for Pontypool and for its Penny Post. In response to these complaints Charles Rideout, who had recently taken over from Woodcock as surveyor, proposed a Penny Post between

---

[1] Archer, Blakely and Jones 1987, 62-3

Caerleon and Pontypool: the average number of letters a day was only four, but the post could be set up without any additional cost, since Caerleon was on the existing route to and from Pontypool. It was simply a modest expansion of the existing Newport-Pontypool Penny Post.

Other evidence points to a later date for the establishment of the Abergavenny Penny Post. Pigot's *Directory* for 1822 only mentions a post to and from Newport, and the earliest known example of a Pontypool Penny Post mark under Abergavenny is from 1831: it would be surprising (although not impossible) if not a single item had survived from its first fifteen years.

On balance it is most likely that the Abergavenny Penny Post was set up in 1827 as one of the enhancements that followed the establishment of a mail coach in April 1827 between Newport and Abergavenny via Pontypool (see below). The following December Rideout reported that a Penny Post had been set up at Pontypool and was likely to produce a gross revenue of over £300 and a profit of nearly £200. He recommended a salary of £20 *p.a.* for the Receiver and 10s per week for a letter carrier. There were to be two free deliveries a day, at 9.00 a.m. and at 3.40 p.m., presumably for the letters from Abergavenny and Newport respectively. This was the first official free delivery to be granted to a community in the valleys: Merthyr, with a population double that of Pontypool, had to wait until 1837. In 1853, when he came to review the rural posts around Pontypool, Trollope recorded a walk that covered various places in and around the town, and this very probably represents the delivery that was set up in 1827. It started on the north side of the town (Penygarn, Trevethin) and then continued through Sowhill to the south-west, and terminated at Lower and Upper Race.

The figure that Rideout indicates as the likely revenue from the Penny Post is far greater than the amount that the Abergavenny post is known to have yielded a few years later and much closer to that of the Newport post. It must represent his estimate for the combined income of both Penny Posts. The actual income for the year May 1835 to May

1836 as given by Rideout in a later report[2] confirms the accuracy of his predictions:

Penny Post revenue at Pontypool, May 1835-May 1836			
	Sent and received	Pence added to General Post	Total
Abergavenny-Pontypool	£44.10.5	2,465d	£54.15.10
Caerleon-Pontypool	£5.9.0	3d	£5.9.3
Newport-Pontypool	£225.8.5	19,254d	£305.2.11

The number of letters between Newport and Pontypool in a year can be derived from the second column. If all the letters had originated from or were addressed to places beyond Newport, the total number of letters (at 1d each) would have been 19,254 , or about 370 a week; if all were purely local within the Penny Post area (at 2d each), then the number would have been about 185.[3] Of course the letters would have fallen into both categories but in what proportion it is impossible to say, so the number of letters received at Pontypool each week at this date lay somewhere between 185 and 370.

Although there is some uncertainty over the date of the establishment of the Abergavenny Penny Post, it is certain that from 1827 Pontypool was served by these two Penny Posts until the introduction of uniform 4d postage in 1839. It probably met the requirements of the commercial classes well enough. Admittedly they had to pay an additional 1d on all their incoming letters, whereas if the town had been made a post town in its own right this would not have been required. But on the other hand, under a Penny Post the cost of a letter to Newport was only 2d, whilst under the General Post it would have been 4d. Pontypool probably preferred to pay the extra 1d on its more distant correspondence in order to keep the cheap rate on the large number of items that went to Newport. Aberdare was placed in a similar position

---

[2] POST 40/639, dated 12 August 1836. Comparable figures for July 1835-July 1836 are given in 'First report of the Select Committee on Postage' 1838, 478, 492
[3] The charge on a letter that was posted and delivered within a Penny Post area was 2d

in 1834, but their preference, for perfectly good local reasons, was to have the General Post extended to Aberdare rather than to be served by a Penny Post under Merthyr – although as it turned out the Post Office ignored this preference and created a Merthyr-Aberdare Penny Post.

Pigot's 1822 *Directory*, in a very interesting entry, describes the postal arrangements then in force, which were probably much the same as when the Penny Post was established in about 1814. The postmistress was Mary Allgood, also listed as an 'Iron Monger' and, very interestingly, 'tallow chandler and japanned ware manufacturer'. Thomas Allgood had invented the process of japanning sheet metal in the late seventeenth century, although it was only brought to perfection by his son, Edward Allgood, in the 1720s. The craft had remained in the family ever since and Mary Allgood was the widow of 'old Billy Allgood', the great-great-grandson of Thomas Allgood. He died in 1810 and Mary Allgood attempted to carry on the business, but without much success. Their only son, William, had no interest in japanning and by 1813 the business was in decline. Mary devoted herself to the ironmongery, chandlery and postal business until her death in 1822. [4] She is buried in the graveyard of the old Baptist chapel at Penygarn: by a nice touch of irony her grave cannot be identified, thanks to the all-prevalent Japanese knotweed.

Mary Allgood's premises were in Lower Crane Street. The post office was carried on in a house which adjoined a larger building, used in Billy Allgood's time as the japanning house.[5] Following her death the post office was taken over by a William Morgan, a draper and grocer, who had premises in what was then known as Caroline Street, now Commercial Street. He was still in office in 1835 and again in 1846.

The 1822 directory entry continues:

> By means of a post caravan, the mails arrive and leave as
> follows: Letters from all parts, by way of Newport, at
> nine in the morning, and dispatched to all parts the same
> way at half-past three in the afternoon.

---

[4] John 1953, 33-41
[5] An engraving of the two buildings by W.H. Greene, dated about 1870, is shown in John 1953, plate 8B. A photograph from the early twentieth century which is said to be of the same building appears in Bradney 1906, 435

A drawing of about 1870, showing the building in Lower Crane Street, Pontypool in which Mary Allgood's post office was located until her death in 1822. The post office is the second building with four windows and a central door. The building below it was used for the manufacture of japanned ware.

The very unusual if not unique term 'post caravan' is probably no more than a grandiloquent description of some sort of horse 'bus which had been adapted by a private contractor for the carriage of mail.[6] The timings given were far from ideal. Both incoming and outgoing mail had to be held at Newport overnight. Mail from London arrived at Newport at about 3.30 p.m., but was not despatched to Pontypool until the following morning, arriving at about 9.00 a.m. Similarly, in the opposite direction, outgoing mail was despatched at 3.30 p.m. and reached Newport at about 5.30 p.m. It was then held there until the following morning before it continued its way at 9.30 a.m. from Newport. All the main mails – London, North, and West – were routed through Newport and so all were subject to this delay. The connection with the westbound mail was even worse, since Pontypool's mail for Cardiff, Swansea and Ireland was held at Newport for nearly 24 hours until it could be put on the down coach at 3.30 p.m. A similar delay affected incoming mail from the west. In July 1822 (after the publication of Pigot's *Directory* for the year) the mail service from London and the rest of England to south Wales was speeded up. Arrival at Newport was now two hours earlier and departure was two hours later, but this simply meant that Pontypool mail was held two hours longer at Newport.

Newport-Pontypool-Abergavenny mail coach

In September 1823 a proposal was made by a un-named stage coach operator who offered to carry the mail from Newport to Pontypool and Abergavenny. Might this have been the proprietor of the 'post caravan' seeking to extend his operation, or was it some other contractor who was anxious to secure the mail contract to support an otherwise uneconomic service? In either case, the Post Office decided to take no action. Lord Granville Somerset had let it be known that he did not want the existing horse post between Newport and Usk to be changed, and as one of the M.P.s for Monmouthshire and a younger son of the Duke of Beaufort his opinion carried weight in official circles. Perhaps

---

[6] As one of the meanings of 'caravan', *OED Online* (2001) has 'A covered carriage or cart: in the 17-18th c. applied to a private or public covered vehicle carrying passengers or a company of people together ...; hence early in 19th c. to a third class 'covered carriage' on a railway ... ' It was clearly a fairly low-status vehicle. The term 'caravan' was also applied to the horse-drawn passenger coach that operated on the Sirhowy Railway between Tredegar and Newport from 1822

he feared that if a mail coach was routed through Pontypool, Usk would no longer have the direct horse post from Newport which had been set up in 1814, but would have to make do with a messenger (possibly on foot) meeting this coach at Pontypool.

However, a few years later, on 5 April 1827, a mail coach was started between Newport, Pontypool and Abergavenny. The timings were a distinct improvement on what had gone before: departure from Newport was immediately after the arrival of the mail from Bristol which meant that Pontypool received its London letters at 3.40 p.m. instead of at 9.00 a.m. the following day. There was then plenty of time overnight to reply by return of post, since the departure from Pontypool was now at about 9.00 a.m., instead of at 3.30 p.m., and this connected with the up coach which left Newport at about 11.30 a.m.

Rideout also proposed to take advantage of the new mail coach to make savings to existing posts. Previously there had been a foot post from Newport to Caerleon and a horse post from Newport to Usk, and mail between Newport and Abergavenny had gone via Usk and Raglan. Under his new arrangements, mail from Newport to Caerleon would go by the mail coach, with a connecting horse post from Caerleon to Usk. Costs would be saved and Rideout claimed that this would produce better outward connections to Bristol, even though the inward arrival time would be later. The Penny Post from Abergavenny to Pontypool was also probably part of the new arrangements, as described above.

Until 1836 all Pontypool's mail was routed through Newport. Very conveniently, the up and down coaches between Bristol and Milford arrived at Newport within little more than an hour of each other, so the mails for Pontypool from both directions could be carried on the same Abergavenny coach. But then, in 1836, a new turnpike road was opened across south Pembrokeshire and the Post Office packets were transferred from Milford Haven to Hobbs Point (at Pembroke Dock). The mail coaches were diverted to Hobbs Point from 11 October but by way of advance preparation the timings of the coaches to and from Milford were speeded up from 5 July 1836.[7] There was now a difference of nearly three hours between the arrival of the up and down mails at Newport.

---

[7] *The Cambrian* 23 April, 15 October 1836

William Needham, manager of the of Varteg ironworks, got wind of the proposals in May and expressed his concern that they would result in Pontypool's London mail being held at Newport for over two hours until the arrival of the up mail from west Wales. This might not seem terribly serious, but it would result in the mail arriving at Pontypool about two hours later. In turn this would allow two hours less to answer by return of post and it would require working later into the evening in order to do so. 'Correspondence between this neighbourhood and the South Wales coast is for the most part wholly unimportant', wrote Needham, and he urged that the Abergavenny coach should not be delayed by having to wait for the up coach from Swansea. By 'the South Wales coast' he presumably meant places such as Swansea and Neath: correspondence between the ironmasters and coal owners of the Eastern Valley and their agents in Newport was extensive and important. The Penny Post revenue for the year ending May 1836 cited above bears this out.

The new timings were duly implemented at the beginning of July and Needham wrote to the Post Office again on 16 July to complain that, as he had feared, the London letters were being delayed at Newport. In order to avoid this he suggested that Pontypool's London and North mails should be routed through Raglan and Usk. This would give an arrival time of 1.30 p.m., over two hours earlier than before the recent changes. It would also avoid the delays in crossing the Severn to which the Bristol mail was subject and would give Pontypool a level of service similar to that enjoyed by Merthyr since 1824. Rideout was asked to investigate the idea. He accepted the benefits that it would bring and estimated the costs at £100 *p.a.* The Postmaster General approved in September 1836 and the new horse post from Raglan to Usk and Pontypool was started probably later the same year.

The Western mail continued to reach Pontypool from Newport on the Abergavenny mail coach. From 1845 its arrival in Newport had been much speeded up by an earlier departure from Bristol following the use of the Great Western Railway for the mail from London. The Western mail now arrived at Newport at 6.53 a.m. The coach to Pontypool and Abergavenny was re-timed to leave Newport at 7.30 a.m. with an arrival at Pontypool at about 9.30 a.m. In the opposite direction the up mail left Pontypool at about 3.00 p.m. to connect with a departure from Newport at 5.37 p.m.

In November 1848 the operators of the mail coach reversed this pattern, with the coach now arriving in Newport in the morning at 10.15 and leaving at 2.15 in the afternoon This was intended to meet the requirements of the travelling public (perhaps to allow a day-time journey to Bristol). The initial response of the Post Office was that Pontypool should be served by means of a mail cart, running to the original times. For some reason this seems never to have been implemented, because when the coach started running to its new times from 20 November the times of the Pontypool and Abergavenny mails to and from Newport saw a corresponding alteration.[8] This meant that Pontypool did not get its Western, Irish and south Wales mails until about 4.00 p.m., over six hours later than before. It is to be hoped that Needham's remarks about the comparative unimportance of this mail were well founded.

Almost certainly associated with this change in some way, since it took effect on the same date as the new schedules were introduced, 20 November 1848, Pontypool was made a post town. It had already, in 1845, been promoted to a Sub Office and Money Order Office. At the same time plans were in hand to build a new post office. The local correspondent of the *Monmouthshire Merlin* reported that 'it is contemplated to erect a Post Office with a reading room adjoining. Such a building, we need not state, would be a desideratum'.[9]

The opening of the South Wales Railway in 1850 made no immediate difference so far as Pontypool was concerned. The mails from London and Bristol now arrived in Newport at 5.20 a.m. but were still not despatched up the valley to Abergavenny until 2.15 p.m. This suggests that the London mail and the North mail both continued to be routed via Abergavenny as before, with only the Western mail transferring to rail at this stage. However, following the opening of the line from Gloucester to Chepstow it is clear that Pontypool did benefit from the improved timings to which this led. From November 1851 the down London mail arrived in Newport at 4.10 a.m. and the coach for

---

[8] *Monmouthshire Merlin* 18,25 November 1848
[9] *Monmouthshire Merlin* 26 August 1848

Pontypool and Abergavenny left at 5.30 a.m. This would have given an arrival in Pontypool by about 7.30 a.m. In the up direction the mail left Pontypool at about 5.00 or 5.30 p.m. in time to connect with the 8.30 p.m. departure of the up mail train. A further improvement in both directions followed the opening of the Wye bridge in August 1852, with the Abergavenny mail leaving Newport at 4.50 a.m.

The question of replacing the Abergavenny mail coach with a mail cart was again raised by the contractors in July 1852. The Post Office agreed that the contractor could use a mail cart rather than a coach between Newport and Pontypool, but there would be no additional payment. The reason for this is because the Monmouthshire Railway had opened their line from Newport to Pontypool on 30 June 1852. Passengers had immediately abandoned the slow and uncomfortable coaches which ceased to be an economic proposition for their proprietors. However, it took another six months before the mail coach actually ceased to run, probably until the existing contract with the Post Office expired, but in January 1853 the mail cart took over, running the full route from Newport to Pontypool and on to Abergavenny.[10] Road transport of one form or another continued to be used on this route until the very end of the century.

Abersychan, Blaenavon and other new receiving houses

During the 1840s the first receiving houses in the Pontypool district were opened. The first of these was in 1845 when an official post was set up to Abersychan, Pontnewynydd 'and the large ironworks in this district'. The fact that the phrase 'official post' was used in the minutes, and the very precise estimate of the level of business to be expected, 429 letters a week, suggest that a private post was already in existence, which is more than likely, given that Abersychan was the site of the works of British Iron Company which was established in 1825. The post of receiver at Abersychan at a salary of £5 *p.a.* was approved on 26 June 1845 and a date stamp was issued on 24 July. James Summers was appointed to the post of foot messenger at 14s a week ('Messenger to the Iron Works under Ponty Pool') on 18 July 1845. It looks as

---

[10] Beaver 2005 dates the withdrawal of the mail coach to January 1854, but January 1853 is indicated by POST 35/120, p.591, dated 17 January 1853. There is no notice of the withdrawal of the coach in the *Monmouthshire Merlin* in either January 1853 or 1854

though there was no great rush of applicants for the post of receiver. On 8 December 1845 the post was still vacant and the surveyor was instructed to select a receiver, and the same again on 6 July 1846. By 1853 this post extended as far north as Garndiffaith (which may well have been its terminus from the outset). Additional receiving houses on the route were opened at Pontnewynydd, Garndiffaith and Talywain (all in September 1853). In his review of the Pontypool rural posts Trollope recommended that in order to provide an improved service for Blaenavon the Abersychan post should be extended to Blaenavon and that control of Blaenavon should be transferred from Abergavenny to Pontypool. The Pontypool-Abersychan messenger was to give up his duties as messenger from Pontypool but he was allowed to continue as subpostmaster at Abersychan. In turn, the existing messenger from Abergavenny to Blaenavon (see below) was transferred to the new Abersychan-Blaenavon walk. These changes were implemented on 4 September 1854.

Blaenavon, about five miles north of Pontypool at the head of the valley of the Afon Lwyd, was the site of a most important ironworks. It had been set up in 1789 by Thomas Hill and his partners, all from the Midlands, who had taken the lease of a 12,000-acre, mineral-rich estate from the Earl of Abergavenny. Within ten years the population was over 1,000 and only Cyfarthfa had a greater output but the location was isolated and its only decent link to the outside world was a tramroad running down the valley to join the Monmouthshire Canal at Pontnewynydd: apart from that it had to make do with mountain tracks over the Blorenge to Abergavenny until the turnpike to Pontypool was opened in 1847.

An official receiving house was opened in Blaenavon in 1847, although a private post to and from Abergavenny and run by the iron company was in operation before that.[11] William Davies was the first receiver, the official minutes noting that he was 'now residing at Pontypool'. He was replaced in February 1849 by William Phillips. The office probably opened in December 1847 when the issue of the first date stamp is recorded. It was placed under the control of Abergavenny and

---

[11] According to Robson's *Directory* for 1840 letters from Abergavenny were received at 3.00 p.m. and dispatched at 10.30 a.m. 'from Isaac Morgans, the Company's store'

was served by a foot post from that place, which went through Llanfoist, Govilon and Garnddyrys. It was reported that the town received just over 500 letters a week. It was made a Money Order Office on 8 August 1851. In February 1853 Crawshay Bailey, owner of the Nantyglo and Beaufort ironworks, submitted a request to improve the service by converting the foot post to a horse post or a coach. No action was taken because by this time Trollope was already engaged on his review of the rural posts and he was expected to make recommendations in any case for improving the service to Blaenavon. The review was completed in July 1853 and Trollope proposed that Blaenavon should be served from Pontypool by an extension of the Abersychan post as described above. Control of Blaenavon was transferred to Pontypool in September 1854.

In October 1849 a post was set up to Pontnewydd, where a tinplate works, on the site of an earlier ironworks, had been in existence since 1802. The village lies to the south of Pontypool and now forms part of the new town of Cwmbran. A receiving house was set up at Croesyceiliog, near the end of the route, and a messenger from Pontypool to Pontnewydd was appointed on 14 November 1849 (John Williams), replaced on 2 October 1850 by Timothy Manley. The district served received 248 letters a week. The post appeared in Trollope's review of July 1853, serving the same places.

Receivers and Postmasters at Pontypool			
	Appointed	In office	Resigned
Mary Allgood			1822 (died)
William Morgan		1827, 1846	
Henry Holloway	1847		1854 (died)
William Edwards		1858	

Chapter 13

Maesteg and the Llynfi valley

At the beginning of the nineteenth century the population of the entire parish of Llangynwyd in the upper Llynfi valley was well under a thousand, most of whom were occupied in agriculture. There was hardly any change in the size of the population until 1826 when the Maesteg Iron Company was formed. The first furnace was blown in in 1828 and in the same year the Duffryn Llynvi & Porth Cawl Railway was opened to connect the area to the sea. The effect of this development was dramatic: between 1821 and 1831 the population of the hamlet of Cwmdu increased from 968 to 2,880.[1]

The so-called 'Old Works' were followed in 1831 by a spelter (i.e. zinc) smelting works located at what is now Caerau at the very top of the valley. Another iron-smelting business followed in 1839, the Cambrian Iron Company, which changed its name to the Llynvi Iron Company in 1844. This company built the Corn Stores, a striking building which was originally an engine-house and has now been incorporated into a sports centre; it represents the most important survivor in Maesteg of the industry which led to the town's creation. A visitor in 1836 described the newly created town as beautifully situated in a valley where its inhabitants were engaged in the iron trade, with rows of white cottages, a company shop and two small chapels.

Maesteg is unusual in that it was the only significant iron-smelting site to have been located within the valleys: all the other important ferrous industry centres were located on the northern or, to a lesser extent, the southern outcrop. Similarly, it was the only town within the heart of the coalfield to have begun its life as an iron town rather than as a coal

---

[1] The parish of Llangynwyd covered much of the Llynfi valley. As was commonly the case with the large upland parishes of south Wales it was divided into a number of hamlets. The iron industry was located in the hamlet of Cwmdu and this was where the town of Maesteg grew up. For a comprehensive modern history of Maesteg, see Richards 1982

town. Iron remained the dominant industry of the town for about thirty years. The output of the two Maesteg ironworks probably peaked in 1859 with an estimated 46,530 tons. This compared not unfavourably with the 50,000 tons a year that Cyfarthfa was producing at this time, although much lower than the 75,000 tons *p.a.* or more of Dowlais. The industry started to go into decline in the 1870s and finally ceased in 1885. With the decline of iron, the sale coal industry, which had been developing since the 1840s, became the dominant industry in Maesteg. It remained a centre of the industry until almost the very end of deep mining in south Wales.[2]

In the absence of an official post before 1843, the need obviously existed for some sort of private service to meet the requirements of the ironworks, the zinc works and a population in 1841 of over 4,000.[3] A private messenger between Bridgend and Maesteg is known to have been functioning in 1843, and he or his predecessor must almost certainly have been active in the mid 1820s if not before. In January 1843 this messenger presented a bill to the Post Office for 10s. It appears that he was seeking compensation for damage or injury caused by an assault. Maberly noted that ' as this is a private messenger ... I presume your Lordship will not interfere. If the party is assaulted he should apply to the local magistrate for protection'. However, and perhaps rather surprisingly, the Post Office agreed to meet his bill.

A request had already been made in 1842 by a Mr Howell for an official post to be established between Bridgend and Maesteg. This was agreed to but only if the inhabitants undertook to guarantee the costs of £39 5s 8d *p.a.* and the matter dropped for the time being. But then, in August 1843, it was decided that the level of business justified an official post, since the number of letters to Maesteg was well over 100 per week. A handstamp was issued on 8 September. On 13 October David Thomas was appointed messenger between Bridgend and Maesteg (on the recommendation of Viscount Adare, M.P. for Glamorgan at the time and later to succeed his father as 3rd earl of Dunraven) and on 16 October Miss A.H. Ballard was appointed

---

[2] For a history of the Maesteg iron industry, see Lewis 2001
[3] This was the population of the parish of Llangynwyd as a whole. The population of Cwmdu alone was 2,880

receiver. Thomas was to be paid £31 5s 8d a year, or 12s a week, which was above average, but for this he had not only to carry the letters to and from Bridgend but also to deliver in the villages which were growing up in the lower part of the valley. Of these the most important was Aberkenfig which had developed as a result of the establishment of the nearby Tondu ironworks in 1836. Delivery within Maesteg itself was probably the responsibility of the receiver.

Business at Maesteg obviously flourished: Miss Ballard was originally appointed at a salary of £4 *p.a.* By 1846 this had been increased to £5 *p.a.* and in December of that year it was increased again to £6 *p.a.* It is explicitly stated in the Minutes that these salary increases reflected a growth in the level of business. The effects of this can also be seen in that by 1848 Miss Ballard had taken on Margaret Llewellyn as an assistant.[4]

In October 1850 further improvements were agreed. The messenger was given a big increase in pay to 21s per week, but for this he was to carry the bags on horseback at 6 m.p.h. instead of on foot. The increase was presumably intended to cover his costs in acquiring and maintaining a horse. At the same time a receiving house was set up at Aberkenfig and a letter carrier appointed to carry out a free delivery within the village, thus relieving the Maesteg postman of this duty. In Maesteg Miss Ballard was given an allowance of £3 4s *p.a.* to cover the costs of delivery within the town, although her basic salary was put back to its original £4 *p.a.* Two months later she married the Revd David Phillips, the minister of Tabor Calvinistic Methodist chapel. On his marriage he took over as receiver, in name at least, even if, as is most likely, his new wife continued to run the office as before. The couple remained in Maesteg until moving to Swansea in 1873. Phillips later held high office within his denomination as President of its General Assembly in 1885.[5]

In 1852 Trollope turned his attention to the Llynfi valley and as a result of his investigations a number of changes were made. A letter carrier

---

[4] The marriage of Margaret Llewellyn, described as 'assistant to Miss Ballard, postmistress, Maesteg' was announced in *The Cambrian* 10 November 1848

[5] The marriage was announced in *The Cambrian* 13 December 1850. For a brief summary of Phillips' life, including his holding the office of postmaster, see *Dictionary of Welsh Biography*, 757-8

was appointed to deliver letters within Maesteg at 6s per week, and the sub-postmaster's allowance was consequently reduced to £1 *p.a.* The mounted messenger from Bridgend continued to receive 21s per week, but he was expected 'to deliver letters from ⅓ mile below the post office southwards, to the Spelter Works northwards including the houses on the turnpike road to the east'. Deliveries in the centre of the town were left to the letter carrier. The Bridgend messenger was probably Thomas Lewis, whose death occurred in 1856: he was described as 'letter carrier from Bridgend to Maesteg' and was said to have been 'greatly beloved for his uprightness and affability of conduct'.[6]

A further receiving house was opened in 1853 at Blackmill, at the junction of the Ogwr Fach and Ogwr Fawr valleys in an area where coal was starting to be worked. Another was started at Caerau in March 1855 with the interesting name of 'Spelter Works' taken from the zinc works that had been set up there in 1831. This district continued to be known as 'Spelter' or 'Spelters' until the end of the nineteenth century when the present name of Caerau became current. The first sub-postmaster was Evan Aubrey, a grocer and draper.[7] All these arrangements remained under the control of the Postmaster of Bridgend, and in fact Maesteg did not become a post town until 1991.

---

[6] *The Cambrian* 18 July 1856
[7] The death of Aubrey's wife, aged 34, leaving him with five young children is recorded in *The Cambrian* 25 May 1855

Chapter 14

Afan valley

Of all the coalfield valleys of south Wales, the Afan valley is one of the least affected by industrialisation. At the lower end, a couple of miles inland from Port Talbot, the valley floor widens out at Cwmavon and offers a suitable site for industrial development, but further inland the steep sides and narrow V-profile of the valley discouraged development. In the tributary Corrwg valley the opening of a railway in 1865 led fairly intensive coal mining and further collieries were sunk in the upper part of the Afan valley after railways were opened from Maesteg in 1879 and Port Talbot in 1885-90. Even so the Afan valley escaped the continuous ribbon development to be found in the valleys further east and today its scenery is by far the most attractive of all the coalfield valleys.

At the point where the river Afan breaks through the hills to flow into Swansea Bay the small medieval borough of Aberavon was established. The name 'Port Talbot' was first applied to the dock at the mouth of the river which opened in 1837. A new settlement grew up around the dock and when the railway opened in 1850 the station was named 'Port Talbot'. This in turn encouraged the use of the name in general for an urbanised area that came to embrace Aberavon, Margam and Taibach.

The mail coach passed through Aberavon but the town was not considered important enough to merit a post office, although there was a receiving house in the town and another at Taibach, a mile or so the south. In 1829 an official post office was set up at Taibach which by this time was becoming the focus for a certain amount of industrial activity and was fairly close to Margam Abbey, the seat of the wealthy and influential Talbot family who were one of the largest landowners in the county. Taibach remained the only post office in the district until the opening of a sub-office at Aberavon in 1843, and even after the opening of the railway it continued to be the principal office in the district until 1885, even though Aberavon was considerably closer to the station.

A mile or so up the valley from Aberavon the metallurgical centre of Cwmavon developed in the early nineteenth century. Coal and iron ore had previously been worked on a small scale but the first large industrial undertaking was the Cwmavon ironworks of 1819. The same company added a tinplate works in 1825 and a copper smelter in 1839. In 1841 the Cwmavon company merged with the 'Governor and Company of Copper Miners in England', a long established undertaking commonly known as the English Copper Company. In the mean time, a little further up the valley, John Reynolds had established the Oakwood ironworks at Pontrhydyfen in 1825 and this too came into the hands of the English Copper Company in 1841. Reynolds was responsible for the construction of the massive aqueduct which still dominates the village. These developments are reflected in the growth of the population of the parish of Lower Michaelstone from 137 in 1801 to 5,421 in 1851, with the main growth occurring in the period 1831-41 and even more so in the following decade.

~~~~~~

A private receiving house in Cwmavon was in existence by 1848, run by John Jones,[1] who can probably be identified with a draper who had premises in Tymaen. There must also have been a private post from Taibach to meet the requirements of the Cwmavon works and perhaps also of Oakwood. Following a petition from the neighbourhood,[2] the private receiving house was upgraded to an official sub-office in 1848, by which time there were said to have been 474 incoming letters a week. Approval for its establishment was given in April 1848, together with a messenger who was to make two journeys a day to and from Taibach, one for the London mail and one for the North mail.

The postal authorities in London seem to have had even greater difficulties than was usual for Welsh place names with 'Cwmavon'. The first handstamp to be issued, on 3 June 1848, was inscribed 'Cromavon'. Obviously someone was convinced that this was the correct name, because in July the surveyor was instructed to select a messenger for

---

1 Prior 2009, 31
[2] Rowlands 1985, 23

the post between Taibach and 'Cromavon'. A second attempt at producing a handstamp resulted in 'Cw-mavon' on 21 June 1848. This too was rejected and finally 'Cwm-Avon' was issued on 24 August.[3]

There is no record of the appointment of the first receiver at Cwmavon, although in all probability it was John Jones who had previously run the private receiving house. In 1854 and again in 1858 a William Button was named as 'postmaster'.[4] The messenger from Taibach who was appointed in 1848 was James Blight. In 1849 he was authorised to perform an additional, but unspecified journey at the request of John Biddulph who at the time was the manager of the works.

Following the opening of the South Wales Railway in 1850 the mails were put off at Port Talbot station. They were then taken to the post office in Taibach for sorting before being delivered. Cwmavon continued to have its two daily deliveries, with the same messenger delivering to both Aberavon and Cwmavon (an arrangement which had probably existed since 1848). In 1856 it was claimed that because letters had first to be taken to Taibach, the North mail, which arrived at the station at about 10.00 a.m., was not being delivered in Cwmavon until 4.00 p.m. – 'and further than that, when the letters did arrive at their destination, their delivery was so uncertain, that parties had to send to the receiving house for them, or wait till the subsequent morning'.[5]

The only other receiving house to be opened in the valley within the period covered by this work was at Pontrhydyfen in June 1857. This was the site of the Oakwood ironworks and of some small coal mines.

Receivers at Cwmavon			
	Appointed	**In office**	**Resigned**
John Jones	1848		
William Button		1854,1858	

---

[3] Broomfield 1996, 77
[4] *The Cambrian*, 14 April 1854, 12 February 1858, in both cases announcing the birth of a child to the wife of Mr Button
[5] *The Cambrian*, 25 January 1856

Chapter 15

Swansea valley

Until it lost its place to Cardiff in the mid nineteenth century, Swansea was generally regarded as the leading town of south Wales. It was both a centre of polite society and one of the seminal sites of the industrial revolution. By Tudor times, and probably before that, it had a healthy export trade in coal and in the eighteenth century it became a centre for smelting copper and other nonferrous metals. Its position on the post road through south Wales to Milford Haven ensured that it had a regular post to and from London at least as early as 1653.

The valley of the river Tawe, running down to Swansea from the foothills of the Brecon Beacons, formed a hinterland which remained predominantly agricultural until well into the nineteenth century. The area lay almost entirely within the coalfield but until the Swansea Canal opened in 1798 commercial exploitation of the reserves of iron and coal was not economically viable; however, they did lead to the development of a number of small ironworks at a fairly early date. A furnace was set up at Ynyscedwyn, towards the top of the valley, possibly as early as 1612. It was converted from charcoal to coke in 1780 and by the 1790s the output was about 800 tons a year (out of a total annual output for south Wales at this period of 12,500 tons). At Ynyspenllwch, near Clydach, an iron forge was established in 1647 and a tinplate works in 1753. However, these were hardly major industries, for in 1801 the population of the entire Swansea valley (apart from the heavily developed area immediately north of Swansea) was less than 2,000. Following the opening of the Swansea Canal a number of sale coal collieries, still small by later standards, opened the length of the valley. Ynyscedwyn expanded and new ironworks were opened at Abercrave (1825), Pontardawe (1835) and Ystalyfera (1838). Other, less successful ironworks were established, many of them in an attempt to take advantage of the use of the hot-blast method of smelting iron ore with anthracite which had been perfected at Ynyscedwyn in 1836.

The output of coal and iron from the valley never approached that of the Merthyr, Cynon or Rhondda valleys. Similarly, the population was never as great as that of the valleys further east. Even so, it showed a steady increase throughout the first half of the nineteenth century with a particularly high rate of growth in the 1830s and 1840s in the period when the iron industry was expanding at Pontardawe, Ynyscedwyn and Ystalyfera.

Private posts

Although there is no evidence for it, it is probably safe to assume that a private post up the valley existed by the end of eighteenth century. Even if the industrial undertakings were only small, there were land-owning families who would expect a regular post, including the Goughs of Ynyscedwyn and the Herberts and Lloyds of Plâs Cilybebyll. The first clear reference to any sort of service is in 1824, when the Ynyscedwyn Iron Company applied to the Postmaster General for an official post to be established from Swansea to Ynyscedwyn. It is surely significant that shortly before this, in 1820, the ironworks had become the property of George Crane, an expansionist ironmaster. He undertook a programme of enlargement and modernisation of the works, and under his enthusiastic management they were to be the scene of innovative attempts at using the local anthracite for the purposes of iron-smelting.

According to figures submitted by Crane in his initial submission to Charles Rideout (11 June 1824), the annual expenditure on postage by users in the valley amounted to £70 4s. Of these the two largest were the Ynyscedwyn Iron Company itself (£20 16s) and John Christie, a major shipper of coal from his Gwaun Clawdd colliery near Abercrave (£18 4s). The balance was contributed by the Abercrave Iron & Coal Company, two collieries in the tributary valley of Cwm Twrch, the landowner Fleming Gough, and the Revd T. Price of Pontardawe. However, on closer examination it became clear to Rideout that the income from a post up the valley would not be as great as might at first sight appear. Whilst there were indeed a good number of items addressed to companies and individuals in the Swansea valley (81 newspapers and 348 letters in a sample month, yielding a total postage income of £13 18s 7d), many of these was in reality going no further than to the offices of the coal and iron companies in Swansea and not to the collieries and ironworks themselves. The remaining items would not generate enough revenue to cover the costs.

> ... from the circumstances of the parties who have extensive collieries in the Vale having Counting Houses at Swansea where their Lrs. are delivered, & which must be a matter of great convenience to their concerns, the post wd therefore only have to convey such communications as they might wish to make with the works &c at Tawe ... Mr Rideout has however assisted the Parties in making a private arrangement for the delivery of their Lrs from the Post Town for one year, and at the end of that time it will be seen whether the correspondence will justify a Post office Establishment.

The above was minuted on 24 August 1824 and Rideout had obviously been as good as his word, for a private mail cart or something similar was already in existence by this date, for *The Cambrian* of 21 August 1824 reported the 'recent establishment of a post-office conveyance from Swansea to the Lamb-and-Flag'. It was not really, of course, a Post Office service, but was paid for by Crane. The terminology goes to show that in the popular mind at this time the distinction between official posts and private posts was not always clear. The mail cart operated three days a week, on Monday, Wednesday and Friday. In the morning it left Swansea at 7.00 a.m., the mail coach from London having come in at about 10.30 p.m. the previous night, and reached the *Lamb and Flag*, a public house at the northern end of the Swansea Canal near the village of Abercrave, at 10.30 a.m. In the opposite direction it left Abercrave at 4.30 p.m. and was due back in Swansea at 8.00 p.m., well in time for the up mail coach due in at midnight.

Presumably these arrangements met the requirements of the coal owners and ironmasters of the Swansea valley, for the service continued after the first year and the question of an official post did not re-appear. In 1830 the three-day post was increased to a daily frequency but by this time the northern terminus had been cut back from Abercrave to Ynyscedwyn.[1] Probably none of the coal owners to the north of Ynyscedwyn were willing to contribute to the costs.

Neither in 1824 nor in 1830 is any indication given as to the identity of the contractor, but in about 1838 the contract was acquired by John Spencer who continued to operate it for over thirty years until he retired in 1871, at first as a private post but from 1848 as an official Post Office contract. He passed into local folk lore along with characters such as Shôn Gwaun Adda of Dinas who carried on a very similar operation in the Rhondda at the same period. Well before the establishment of the official post up the valley Spencer appears to have run a well planned service, based very much on the official model. The following tribute, published on his retirement, describes the way in which he carried out his business:

> The name of Spencer, in connection with the mails in this
> Valley, in consequence of his great length of service, has
> become a bye-word and a proverb, and he himself seems
> quite an institution in which all classes seem to have an
> interest of no ordinary kind. For thirty-two long years has
> he carried the mails up and down the Valley. About
> twenty-two years ago the first post-office here was
> opened, but for ten years before that time Spencer had
> been the regular letter-carrier for the whole Valley. It
> appears that at that time all letters were entrusted to
> Spencer in a bag loosely, and that by the permission of
> the post-office authorities, he was authorised to charge
> two pence for every letter. This money was equally
> divided between Spencer and the owners of the receiving
> houses selected by him for depositing the letters on his
> route from Swansea to Ystradgynlais, a distance of about
> fifteen miles or thereabouts. When post-offices were
> subsequently opened at the principal villages along the
> route, those receiving-houses, for the most part, became
> the authorised post-offices ... For many years Spencer
> travelled up and down the Valley by night, leaving
> Swansea after the arrival of the mail at about 5 or 6

---

[1] *The Cambrian* 9 January 1830

o'clock p.m., and returning in time for the despatch of the
North mail, at 6 o'clock next morning ... [2]

The village of Cwmtwrch, which lay in a tributary valley to the
Swansea valley, also had a private messenger at the same period, who
was employed by the inhabitants to carry their letters to and from the
post office in Swansea. In 1841 Thomas Jackson submitted a request
for the establishment of a regular post and the appointment of an
official messenger. Hardly surprisingly he was told that the level of
correspondence did not justify an official post and that his request
could not be met unless the parties concerned gave a guarantee that
they would meet the costs, which were estimated to be £51 14s 4d *p.a.*
The wording of the official response seems to show that Jackson had
commented unfavourably on the private messenger, implying that an
official messenger would be more dependable. The Post Office did not
take the bait.

> I submit Mr Jackson may be informed that the party to
> whom he alludes is employed by the Inhabitants of Cwm
> Twrch and not by the Post Office, they can discharge him
> and employ any other party they prefer to call for the
> letters at the Post Office in Swansea .

No more was heard of this request. Cwmtwrch did not have a post
office until 1882, although by 1853 it had a daily delivery of letters
from Ystalyfera.

Official post and sub-offices

By 1847 the demand for an official post up the Swansea valley was
once again making itself felt. The ironworks at Ynyscedwyn were at
their peak and further ironworks had recently been opened at
Ystalyfera. The population of the valley was approaching 9,000 and
growing rapidly. In that year a petition was presented to the Postmaster
General by 'the principal iron-masters, colliery proprietors, copper
smelters, merchants, &c., interested in the trade of the district'
requesting a daily mail coach between Swansea and Ystradgynlais.
'[T]he want of a daily communication up the Swansea Valley, from the
great increase of traffic which is constantly taking place, has long been

---

[2] *Ibid* 18 August 1871

felt, and merits the immediate attention of the postal authorities.[3] Perhaps by the call for a 'daily' post means that the petitioners wanted a seven-days-a-week service and Spencer was only operating on weekdays; or it could mean that Spencer had reduced the frequency of the service from the six days a week of 1830. In any case, an official post was approved on 24 April 1848 and the first three post offices in the Swansea valley, at Clydach, Pontardawe and Ynyscedwyn, were opened. By this time about 500 letters a week were being received in the valley, which was well over the figure of 100 per week that was normally required by the Post Office at this time for an official post to be set up. Handstamps were issued on 4 July 1848 and receivers were appointed on 1 August. Presumably business commenced shortly afterwards. Of the three handstamps, two contained errors: Pontardawe was spelt 'Ponterdawe' and Clydach became 'Claydock'. But they seem to have been accepted and replacements were not issued for several years; even then the new Pontardawe mark was still incorrect, reading 'Pontradawe'.[4]

Spencer obtained the Post Office contract to carry the mails up the valley and marked his success by investing in a new coach which started to run in July 1849.[5] Perhaps he had built the coach himself, for in the 1851 census enumerators' returns he was described as a coach builder; he lived outside Swansea at Weig Fawr farm. On 17 July it was minuted officially that he was to be allowed to run a coach with bags under the usual privileges, which was not the same thing as a mail coach as the term is normally understood. Spencer continued to hold the contract until 1871, but by 1864 his new coach had become decidedly the worse for wear and complaints were made about the way in which the Swansea valley mails were carried in 'about the slowest and most antiquated conveyance which the parsimony of the General Post Office can obtain'.[6] One is reminded of the reactions in Merthyr to the mail cart which was put on to serve that town in 1850.

The receiving house at Ynyscedwyn was very soon discontinued. Authority was given on 26 June 1849 for it to be replaced by a new one

---

[3] *Ibid* 29 October 1847
[4] Broomfield 1996, 77
[5] *The Cambrian* 13 July 1849
[6] *Ibid* 29 April 1864

in Ystradgynlais. However, the name of the receiver remained the same, so perhaps it was simply that the name of the office was changed rather than that a totally new office was set up. Pontardawe and Ystradgynlais both attained Money Order Office status in March 1851. A fourth office at Ystalyfera was approved on 3 November 1851 and opened in February 1852.[7] The receivers at all four offices received an allowance for making a local delivery of letters. In December 1852 there were daily deliveries in Clydach and Ynyspenllwch (from Clydach); Pontardawe; Ystalyfera, Ponttwrch and Cwmtwrch (from Ystalyfera); and Ystradgynlais, 'the delivery to extend down to the works' (i.e. Ynyscedwyn ironworks). By November 1859 a sub-office had been opened at Abercrave, with deliveries to the village itself six days a week and extended as far as to Craig-y-Nos three days a week.

With the opening of the South Wales Railway to Swansea in 1850 the times of the Swansea valley mail cart were completely re-cast to connect with the mail trains. Spencer was no longer required to travel the valley by dark. He now left Swansea at 9.00 a.m. and reached Ystradgynlais at 11.30 a.m. He returned from there at 1.15 p.m. and was due back in Swansea at 3.45 p.m. Spencer retired in 1871, but the pattern remained much the same until the 1880s when a second North mail despatch was instituted later in the morning and, some time after that, the Swansea Vale Railway finally started to be used to carry mail.

---

[7] *The Cambrian* 13 February 1852, says 'this week', which can presumably be accepted as correct; the first handstamp was issued on 11 December 1851 to 'Ystralafera' (Broomfield 1996, 78)

Appendix

Private posts

The following is a list of privately organised posts which are known to have operated within the south Wales valleys. Much of the information is derived from the Woodcock papers now in the Gloucester reference library.[1] Shortly after his appointment as surveyor for the western district Charles Woodcock started to compile a detailed summary, town by town, of the postal arrangements within his district. The date of compilation can be put to 1787 or later by the fact that Cardiff is shown as having a six-day mail coach whereas Brecon only had a three-day postboy: the mail coach to Hubberstone was transferred from the northern to the southern route in August 1787. Unfortunately Woodcock only got as far as Cardigan, and even then some of the earlier towns have blank pages. Consequently only the entries for Brecon and Cardiff contribute to this list.

The other posts which are listed are known from documents in the Post Office Archive, where the existence of a pre-existing private post is mentioned at the time that an official post was established, and from contemporary directories. Almost certainly there must have been many other private posts within the valleys for which no firm evidence has survived

Abercynon to Aberdare. Ceased 1834

Abergavenny to Blaenavon (in existence by 1840)
Abergavenny to Ebbw Vale. Commenced 1824 (carried on mail coach as a private bag)

Brecon to Merthyr Tydfil. In existence 1766; ceased 1787
Brecon to Merthyr Tydfil (by carrier). In existence 1795

Bridgend to Maesteg. Ceased 1843

---

[1] 'Woodcock Papers' 1992.

Cardiff to Aberdare. In existence c1790

Cardiff to Aberpergwm (Neath valley). In existence c1790; probably carried to Neath on mail coach as a private bag

Cardiff to Caerphilly. In existence c1790; ceased 1839

Cardiff to Energlyn. In existence c1790

Cardiff to Gelligaer. In existence c1790

Cardiff to Llanbradach. In existence c1790

(The four destinations above may all have served by the same post)

Cardiff to Llanfabon. In existence c1790

Cardiff to Llantrissant and Castellau. In existence c1790

Cardiff to Llanwonno. In existence c1790

Cardiff to Merthyr Tydfil. Possibly in existence 1742

Cardiff to Merthyr Tydfil. Commenced 1786; ceased 1804

Cardiff to Newbridge [i.e. Pontypridd]. In existence c1790.

Glynneath to Banwen. In existence 1848

Newbridge [i.e. Pontypridd] to Aberdare. Commenced c1815; replaced by Abercynon-Aberdare 1821

Newbridge [i.e. Pontypridd] to Cymmer and Dinas. Ceased 1850

Newport to Abercarn. In existence 1814 (and probably much earlier)

Newport to Pontypool. In existence 1807; ceased c1816

Newport to Sirhowy, Ebbw Vale and Nantyglo. Commenced early 1800s; ceased 1839

Newport to Tredegar. Commenced early 1800s; ceased 1839

Swansea to Ynyscedwyn. Commenced 1824; ceased 1848

Sources and bibliography

MANUSCRIPT MATERIAL

Royal Mail Archive

Records on Conveyance of Mails by Road, Inland Service. POST 10

Records on conveyance of mail by Railways. POST 11

Revision of rural posts in England, Wales, Scotland and Ireland, case files. POST 14/6

Registered Files, Minuted Papers (England and Wales). POST 30

Minutes between the Secretary and Postmaster General: Volumes. POST 35

Postmaster General's Reports: Documents. POST 40

Postmaster General's Reports. POST 42

Appointments taken from minutes. POST 58/02

Glamorgan Record Office (Glam RO)

Dowlais Letter Books: Main Series (incoming letters 1792-1917). DG/A/1

Dowlais Letter Books: Outward letters (1782-1794). DG/A/2

Dowlais London House Letter Books (1824-1900). DG/A/3

Dowlais Cardiff Agency letters (1819-1893). DG/A/4

Gloucester Reference Library

Woodcock papers. B447/48237GS (entries relating to Wales transcribed in the *Welsh Philatelic Society Newsletter*, see below)

Gwent Record Office (GRO)

The Letterbook of Richard Crawshay. D.2.162 (calendared by Evans 1990; see below)

Ebbw Vale Company. Letter Book. D.2472.2

Ebbw Vale Company. Journal 1814-1815. D.2472.4

National Library of Wales (NLW)

Cyfarthfa papers. GB 0210 CYFHFA

Accounts for Homfray Ironworks. MS 15593E

Letter Book of Richard Hill 1786-1792. MS 15334E

Maybery papers. GB 0210 MAYBERY

West Glamorgan Archive Service

Leslie Evans note books. D/D LE 59

PRINTED MATERIAL

Books

Michael Scott Archer, *The Welsh Post Towns before 1840* (Chichester, 1970)

Michael Scott Archer, Robin Blakely, Geraint Jones, *The Welsh Post Towns before 1840*: *Supplement* (1987)

Michael Atkinson and Colin Baber, *The Growth and Decline of the South Wales Iron Industry 1760-1880* (Cardiff, 1987)

Chris Barber, *Eastern Valley : the Story of Torfaen* (Llanfoist, 1999)

Philip Beale, *A History of the Post in England from the Romans to the Stuarts* (Aldershot, 1998),

Christopher M. Beaver, *The Welsh Mail Coaches* (2005)

J.A. Bradney, *A History of Monmouthshire. Pt.II: The Hundred of Abergavenny* (London, 1906)

Roger W. Broomfield, *The Undated Postmarks of Wales 1840 to 1860* (1996)

Harold Carter & Sandra Wheatley, *Merthyr Tydfil in 1851 : a Study of the Spatial Structure of a Welsh Industrial Town* (Cardiff, 1982),

John Charles, *Pontypridd Historical Handbook* (Pontypridd, 1920)

Cynon Valley History Society, *Cynon Coal* (Aberdare, 2001)

M.J. Daunton, *Royal Mail: the Post Office since 1840* (London, 1985)

Paul Davis, *Historic Rhondda: an Archaeological and Topographical Survey 8000 BC - AD 1850* (Rhondda, 1989)

Dictionary of Welsh Biography down to 1840 (London, 1959)

John Elliott, *The Industrial Development of the Ebbw Valleys 1780-1914* (Cardiff, 2004),

Kenneth Ellis, *The Post Office in the Eighteenth Century* (London, 1958)

Madeleine Elsas (ed), *Iron in the Making: Dowlais Iron Company Letters* (Cardiff, 1960)

Chris Evans, *'The Labyrinth of Flames': Work and Social Conflict in Early Industrial Merthyr Tydfil* (Cardiff, 1993)

Chris Evans (ed.), *The Letterbook of Richard Crawshay 1788-1797* (Cardiff, 1990) (A calendared edition of GRO D.2.162; see above)

Arthur Gray-Jones, *A History of Ebbw Vale* (1970)

Joseph Gross (ed.), *The Diary of Charles Wood of Cyfarthfa Ironworks, Merthyr Tydfil, 1766-1767* (Cardiff, 2001)

Stephen Hughes, *The Brecon Forest Tramroads: the Archaeology of an Early Railway System* (Aberystwyth, 1990)

Laurence Ince, *The South Wales Iron Industry 1750-1885* (Solihull, 1993)

C.H. James, *What I Remember about Myself and Old Merthyr* (Merthyr Tydfil, 1892)

J.E. Jenkins, *Vaynor: its History and Guide* (Merthyr Tydfil, 1897)

W.D. John, *Pontypool and Usk Japanned Wares* (Newport, 1953)

Alan Vernon Jones, *A Translation from the Welsh of William Bevan's History of Mountain Ash 1896 with additional notes and illustrations* (Aberdare, c1990)

Gareth Elwyn Jones, *The Education of a Nation* (Cardiff, 1997)

Ieuan Gwynedd Jones, *Health, Wealth and Politics in Victorian* Wales (Swansea, 1979)

Oliver Jones, *Early Days of Sirhowy and Tredegar* (Tredegar, 1969)

F. George Kay, *Royal Mail* (London, 1951)

D.A. Lewis, *A Cambrian Adventure: a History of the Iron Industry in Maesteg* (Port Talbot, 2001)

E.D. Lewis, *The Rhondda Valleys* (London, 1959)

B.H. Malkin, *The Scenery, Antiquities and Biography of South Wales* (London, 1802)

'Morien', *History of Pontypridd and Rhondda Valleys* (Pontypridd, 1903)

John A. Owen, *A Short History of the Dowlais Ironworks 1759-1936* (Merthyr Tydfil, 1972)

Don Powell, *Victorian Pontypridd and its Villages* (Cardiff, 1996)

Evan Powell, *History of Tredegar* (Tredegar, 1902)

Brinley Richards, *History of the Llynfi Valley* (Cowbridge, 1982)

Philip Riden, *A Gazetteer of Charcoal-fired Blast Furnaces in Great Britain in use since 1660* (2nd ed., Cardiff, 1993)

John Rowlands, *In the Valley Long Ago* (trans. Graham Hughes) (New York, 1985)

Stephen Rowson and Ian L. Wright, *The Glamorganshire and Aberdare Canals. Volume 1: Merthyr Tydfil & Aberdare to Pontypridd* (Lydney, 2001)

Charlotte Schreiber, *Lady Charlotte Guest: Extracts from her Journal 1833-1852* (London, 1951)

Keith Strange, *Merthyr Tydfil, Iron Metropolis: Life in a Welsh Industrial Town* (Stroud, 2005)

Charles Wilkins, *The History of Merthyr Tydfil* (Merthyr Tydfil, 1867; 2nd ed. 1908)

Charles Wilkins, *The History of the Iron, Steel, Tinplate, and Other Trades of Wales* (Merthyr Tydfil, 1903)

Charles Wilkins, *The South Wales Coal Trade and its Allied Industries: from the Earliest Days to the Present Time* (Cardiff, 1888)

R.M. Willcocks, *England's Postal History to 1840, with Notes on Scotland, Wales and Ireland* (1975)

Gwyn A. Williams, *The Merthyr Rising* (London, 1978; 2nd ed, Cardiff, 1988)

Dissertation

Keith H. Edwards, 'William Williams (b. Newbridge, Glamorgan, 1808 d. Cardiff, 1890), grocer and draper at Brynmawr : his account books and diaries and the development of Brynmawr' (unpublished M.A. thesis, University of Wales College of Cardiff, 1997)

Articles etc.

Len Burland, 'Homfray: an industrial dynasty', *Gwent Local History* 101 (2006), 3-34

Elizabeth Havill, 'William Taitt, 1748-1815', *Trans Cymmrodorion* 1983, 97-114

Hen Gymraes, 'Ebbw Vale in the 1840s: life in the Monmouthshire hills', *Gwent Local History* 84 (1998), 35-49 (extracted from reminiscences originally published in the *South Wales Argus* in 1921)

John B. Hilling, 'Britain's first planned industrial town? The development of Tredegar, 1800-1820', *Gwent Local History* 94 (2003), 55-76

John B. Hilling, 'The migration of people into Tredegar during the nineteenth century', *Gwent Local History* 100 (2006), 19-40

T.M. Hodges, 'Early banking in Cardiff', *Economic History Review* 18 (1948), 84-90

Bill Jones, 'Writing back: Welsh emigrants and their correspondence in the nineteenth century', *North American Journal of Welsh Studies*, 5 (2005), 23-46

Neil Prior, 'Postal services in 'old' Port Talbot', *Welsh Philatelic Society Newsletter* 112 (2009), 26-33

Karen Isaac Rees, 'The Bassetts of Sigginston', *Llantwit Major: Aspects of its History*, 8 (2008), 72-85

R.D. Rees, 'Glamorgan newspapers under the Stamp Acts', *Morgannwg*, 3 (1959), 61-94

Paul Reynolds, 'The Brynmawr-Newbridge tramroad mail coach, 1839', *Welsh Philatelic Society Newsletter*, 35 (1983), 18-21

Paul Reynolds, 'Charles Wilkins of Merthyr Tydfil, postmaster and historian', *Welsh Philatelic Society Newsletter* 65 (1994), 5-9

Paul Reynolds, 'The Merthyr to Abergavenny mail coach', *Welsh Philatelic Society Newsletter* 103 (2006), 16-21

Paul Reynolds, 'The Merthyr to Abergavenny mail coach : a post-script', *Welsh Philatelic Society Newsletter* 105 (2007), 18-20

Gerald Richards, 'Death of a postman', *Welsh Philatelic Society Newsletter* 107 (2008), 24-5

R.O. Roberts, 'Commercial banks in Glamorgan' (unpublished typescript appendix to *Glamorgan County History* v 5; copies deposited in the National Library of Wales, Swansea University library and elsewhere) (c.1980)

R.O. Roberts, 'The operations of the Brecon Old Bank of Wilkins & Co. 1778-1890', *Business History* 1 (1958), 35-51

G.B. Smith, 'Freeling, Sir Francis, first baronet (1764-1836)'; rev. Jean Farrugia, *Oxford Dictionary of National Biography* (London, 2004)

'Some notes on the early postal history of Aberdare', *Welsh Philatelic Society Newsletter* 5 (1973), 27 (Attributed to the late Revd Ifor Parry)

John Wilkins, 'Charles Wilkins, writer, 1830-1913: a biographical note', *Merthyr Historian* 13 (2001), 5-13

John Winstone, 'Reminiscences of old Cardiff', *Report and transactions [of the] Cardiff Naturalists' Society,* 15 (1883), 60-75

'The Woodcock Papers', *Welsh Philatelic Society Newsletter* 59 (1992), 16-18 (Brecon); 60 (1992), 13-15 (Cardiff)

Directories

The Universal British Directory of Trade, Commerce and Manufacture (1791-98)

Pigot and Co.'s London and Provincial New Commercial Directory (1822/23)

James Pigot and Co's National Commercial Directory (1830)

Pigot & Co's National Commercial Directory ... South Wales (1835)

Robson's Commercial Directory of London and the Western Counties ... and South Wales (1840)

Pigot & Co's Royal National & Commercial Directory ... Monmouthshire ... (1842)

Pigot & Co, *Royal National & Commercial Directory ... South Wales* (1844)

Hunt & Co.'s City of Bristol, Newport & Welch Towns Directory (1848)

Hunt & Co.'s Directory & Topography for the Cities of Gloucester & Bristol : and the Towns of Carmarthen [etc.] (1849)

Slater's (late Pigot & Co.) Royal National and Commercial Directory and Topography of the Counties of ... Monmouthshire ... (1850)

Scammell & Co's City of Bristol and South Wales Directory (1852)

Slater's (late Pigot & Co.) Royal National and Commercial Directory and Topography of the Counties of Gloucestershire, Monmouthshire and North and South Wales ...(1858/59)

Newspapers

Aberdare Leader (Aberdare)

Aberdare Times (Aberdare)

The Cambrian (Swansea)

Cardiff & Merthyr Guardian (Cardiff)

Glamorgan, Monmouth & Brecon Gazette. Continued by *Glamorgan, Monmouth & Brecon Gazette and Merthyr Guardian* (Merthyr Tydfil)

Merthyr Guardian (Merthyr Tydfil)

Monmouthshire Merlin (Newport)

The Times (London)

Western Mail (Cardiff)

Parliamentary papers

'First report from the Select Committee on Postage; together with the Minutes of evidence, and Appendix. Appendix 4: Return ... showing the number of letters posted ... for one week ... '. 1837-38 (278) 448

'Second report from the Select Committee on Postage; together with the Minutes of evidence, appendix and index. Appendix E: Return as to numbers of letters and postage'. 1837-38 (658) 469

'Second report from the Select Committee on Postage; together with the Minutes of evidence, appendix and Appendix. Appendix 45: Returns relating to mail coaches, etc'. 1837-38 (278) 657

'Second report from the Select Committee on Postage; together with the Minutes of evidence, appendix and Appendix. Appendix 53: Returns relating to mail carts, etc'. 1837-38 (278) 668

'Return of Number of Stage Coaches used as Mails in England and Wales'. 1843 (602) 327

'Return of Number of Mail Guards in Great Britain and Ireland employed in Post Office, 1837-40'. 1841 Session 1 (431) 381

Index